Penguin Special

Fuel's Paradise: Energy Options for Britain

Peter Chapman was born in Yorkshire in 1944. He was
brought up in London and won an Exhibition to
Emmanuel College, Cambridge, to study Natural Sciences.
After obtaining a first class honours degree in Physics, he
went on to research in electron microscopy at the
Cavendish Laboratory. He subsequently became interested
in the social aspects of science and technology, particularly
with regard to energy use and climate. Leaving Cambridge,
he joined the Open University as a physics lecturer.

After four years of full-time teaching, the energy ideas
begun in Cambridge were developed in the production of a
course on Materials Science and led to the establishment
of the Energy Research Group of which he is director.

Peter Chapman is now a senior lecturer in physics at the
Open University and is well known in the energy field as
the author of many papers and as a critic of conventional
energy policies. He has also made some television
programmes on energy.

Peter Chapman

# Fuel's Paradise
*Energy Options for Britain*

 Penguin Books

Penguin Books Ltd,
Harmondsworth, Middlesex, England
Penguin Books Inc.,
7110 Ambassador Road, Baltimore, Maryland 21207, U.S.A.
Penguin Books Australia Ltd,
Ringwood, Victoria, Australia
Penguin Books Canada Ltd,
41 Steelcase Road West, Markham, Ontario, Canada
Penguin Books (N.Z.) Ltd,
182–190 Wairau Road, Auckland 10, New Zealand

First published 1975
Copyright © Peter Chapman, 1975

Made and printed in Great Britain by
Cox & Wyman Ltd, London, Reading and Fakenham
Set in Monotype Times

This book is sold subject to the condition that
it shall not, by way of trade or otherwise, be lent,
re-sold, hired out, or otherwise circulated without
the publisher's prior consent in any form of
binding or cover other than that in which it is
published and without a similar condition
including this condition being imposed on the
subsequent purchaser

# Contents

Acknowledgements 8

1. A Parable: The Isle of Erg 9
2. Energy Currency 21
3. Why We Need Fuels 31
4. How We Use Fuels 40
5. Where It Goes To 58
6. The Problem of Too Much 70
7. Fuel Supply Efficiency 89
8. Energy Policy 110
9. A Prologue on the Future 120
10. Futures I: Business-as-usual 124
11. Futures II: Technical-fix 155
12. Futures III: Low-growth 186
13. The Energy Shop 209

Appendix: Fuel Use in Transport 220
Suggestions for Further Reading 229
References 231
Index 235

You can fuel all the people some of the time
and
You can fuel some of the people all the time
but
You can not fuel all the people all of the time.

# Acknowledgements

This book is essentially a personal summary of the things I have learnt from discussions with many people over the past four years. I owe most to Mike Hussey for his wisdom in perceiving the world with alarming clarity. I am also indebted to the members of 'Energy 2000' who set out to produce a document similar to this, in particular to Walt Patterson, Gerald Leach, John Price and Amory Lovins, all of whom have taught me a lot. Colleagues at the Open University, especially members of the Energy Research Group, have helped with thoughts and detailed criticisms. I am grateful to Bob and Gary for their care in reading early drafts and to Joan Harrison for translating my manuscript into a typescript. By their understanding and love Ev, Jayne and Linda have had a large role in making this book possible. However, even all these wise people are not responsible for any mistakes which may appear in the following pages. This simply reflects how much more I still have to learn.

# 1 A Parable: The Isle of Erg

## 1

*Next month the electorate of the Isle of Erg go to the polls to decide whether to continue with their isolationism or to open the doors to world trade and, as one opposition spokesman put it, 'take a leap into the twentieth century'. This is the first of a series of special reports on the island economy and covers the background to the impending referendum.*

For the last forty-five years the only trade between this small Mediterranean island and the rest of the world has been the trickle of exchanges approved by the Erg government. All this could change next month if Valar's government is defeated at the polls. Although the trade issue has attracted most attention outside the island it is not an important domestic issue. The real dispute is between the government and the would-be businessmen who find their opportunities restricted by official policy. At the heart of the dispute is the government's housing policy and its controversial xat system.

The last referendum in 1970 produced an overwhelming victory for the government, but since then the opposition party has increased its strength. Many of the new voices raised against the government belong to the two hundred immigrants who have moved in since 1970. These are all people born on the island who left in the 1950s for better prospects abroad. Although this group has injected new life and vigour into the opposition movement they don't seem to have convinced many of the other islanders to vote against the system.

The government has responded to the opposition attacks by

publishing fairly detailed accounts of the problems currently facing the advanced industrial nations in the wake of the oil crisis. Most of the facts are not in dispute, but the government interpretation, mostly the work of Kaycal, the Minister of Industry, is a bone of contention. Kaycal's theory is that the oil crisis has caused the industrial nations to move closer to the Erg type of economy and that in time the Erg system is bound to prove superior. The chief opposition spokesman, Malvegil, has called this 'romantic rubbish' and argues that the oil crisis is only a hiccup in the continued development of the industrial nations. 'Five years from now they will be back on the path to increased production and better standards of living and they will leave the Isle of Erg back in the 1930s,' he claims. Malvegil wants to see the xat system scrapped, the restrictions on housing removed and the country's economy opened up to world trade.

Of these three bones of contention the most serious is the xat system, since this is seen as one of the major obstacles to the business community. The xat system was the invention of Valar, now Prime Minister, and was an extension of the rationing schemes brought in during the 1930s and 1940s. The system gives every man, woman and child on the island a government income. For most people this income is more than enough to live on and, according to Malvegil, removes all incentive to work. Although the guaranteed income does stop people applying for jobs in the industrial sector it doesn't stop them working.

Our next-door neighbour, Galen, has two children and together he and his family have a weekly income of 1,000 kwats. Galen spends two days a week as a taxi-driver. For this he earns an extra 40 kwats. Galen's wife is one of the many men and women who help out at the local school on a voluntary basis. When Galen isn't off with his horse-drawn taxi he is either working in his garden or helping his friend Beren in his cycle business. Over the past month I've seen Galen working in his garden almost daily. Although this doesn't bring him any income directly it does provide his family with about half the food they need and it reduces the rates on his house. When I asked Galen why he bothered with the taxi business he said it was to pay for the horse, which was useful in many other ways, and because he enjoyed going out and meeting people

This type of attitude is common among the people who live in the village. Beren, who lives at the end of the lane, charges only as much for his bicycles and repair work as he needs to buy more materials and keep his forge running.

The xat system works against businesses in two ways. As Galen, Beren and the other villagers testify the combination of a xat income and a large garden certainly provides a basis for a very comfortable life, so no one has any reason to seek employment with anyone else. Even going out to work for a firm would not increase their incomes very much, because the major costs of producing anything on the island are fuel and materials costs. All fuels and materials are produced by government-owned industries; the profits from these industries are used to finance the xat system. The problems were starkly posed by Morgoth, a small manufacturer in Brittu who wants to go into the bicycle business. Morgoth's company already produces most of the ball bearings, chains and cables used in bicycle manufacture. Now he wants to install a lathe, some welding equipment and paint spray to set up a bicycle production line. 'The trouble is that I can't find anyone prepared to come and work the machines. It's a crazy situation, because with just ten workers we could produce more bicycles in a month than the two hundred workshops scattered around the country.' Morgoth claims that this would leave the other hundred and ninety cycle-makers free to make other goods, 'perhaps washing machines or any of the other thousand things not produced'. But even if Morgoth found his ten workers he would still face problems, since at the moment hand-made goods are always cheaper than machine-made. According to Morgoth this is because of the government's monopoly of the fuel and materials industries. 'What other country in the world charges 8,000 kwats for a ton of coal, or 40 kwats for the electricity needed to run a welding set for an hour, and has a workforce employed at an average weekly wage of 100 kwats?'

Morgoth's problems are not the causes but the symptoms of the dispute. I asked Galen whether he would consider going to work for Morgoth's company. He pointed out that even if he worked there full-time and earned 100 kwats a week this would be offset by his loss of income from taxi work and by the increase in his rates as soon as he stopped working his garden so intensively. He also asked me why someone should want to produce more bicycles 'when there are already enough for everyone and the small local workshops could cope with all the necessary repairs'. 'The only reason for producing a hundred new cycles a month is to put people like Beren out of business, and why should I be interested in doing that?' In point of fact Galen had reasons for keeping Beren in business, since by letting him use the taxi for transporting materials Galen had most of his repairs done free.

According to Valar's government, businessmen like Morgoth are trying to destroy the basis of the Erg community by trying to persuade the islanders to consume more. Morgoth is unlikely to persuade Galen and people like him because they are content with their lives and fear the risks associated with change. The fact that most of the islanders cannot be persuaded to consume more has led the opposition to launch attacks on the government policies on housing and trade. Since 1930 housing regulations have forbidden the construction of new homes. The regulations were introduced when building materials were very scarce but have been continued as a measure to control population and hence domestic demand. The restrictions on external trade were also introduced in the 1930s, but not voluntarily. Until 1922 the island had traded, almost exclusively, with Italy. A steady food surplus sold in Italian markets purchased machinery, fuels, raw materials and consumer goods. In 1922 Mussolini introduced his fascist regime in Italy and the socialist government of Erg immediately cut off all diplomatic and trading links. This produced difficulties for two or three years, but by 1925 Erg was obtaining sufficient supplies of fuel and machinery from France to keep its economy ticking over. Four years later the Wall Street crash sent all the industrial countries into the Great Depression and the islanders found that they could not sell their food surplus. By 1930 the shortages of fuel and machine parts on the island had brought its industries to a halt. From that time until the end of the Second World War fifteen years later the island was effectively isolated by the rest of the world. In those fifteen years the detailed rationing scheme became the xat system. The government took over the major industries and conscripted a workforce to man the machines while other nations were conscripting men for armies. By 1945 the Erg community had found its feet and developed a self-sufficient economy. Every five years since then the island has held a referendum to decide whether to continue with its now self-imposed isolation or to re-establish contact with the outside world. Next month is the seventh time that the islanders have been asked to choose and it looks as if they will stick with their system for a while yet.

2

*'If we tried to live the way we used to in London we would need an income four times as big, but we haven't noticed a drop in our standard of living – just a change.' This second report from the Isle of Erg describes what it's like balancing the housekeeping in a unique economy.*

The contrast between London and Joule, the capital of Erg, is obvious as soon as you step out of the aircraft at Joule airport. The main terminal building looks more like an oversized Welsh farmhouse than the centre of communication with the outside world. Made of stone, with only a few small windows, the buildings are in stark contrast to the glass and concrete structures left behind such a short time ago. Most newcomers leave the airport in one of the horse-drawn taxis. These are specially laid on for visitors, the islanders preferring to propel themselves around on bicycles, tricycles – even pedal-driven buses. Although there is nothing like the six-lane carriageway which marks the approach to most of the world's airports the road to town is open and never congested. On one side the houses are close to the road, but on the south side the spacious gardens stretch for forty or fifty yards up to the houses. Most of the houses, like the airport buildings, are made of stone with a single pitched roof, usually covered with glass. But the most striking visual image is the windmill. There are as many windmills in Joule as there are TV aerials in an equivalent-sized London suburb. Normally fixed on the roof, sometimes on a separate tower, sometimes on an old tree, every house has its windmill.

We have been living in a typical island house in a small village just outside Brittu, the second largest town on the island. The house is rented to us on an exchange scheme whereby my salary goes to the Erg government and in return we have the house and the normal xat allowance for our family of four. The house, its garden and animals is an institution of its own. We have been here for three weeks now and are still finding out how to live in the house. By European standards it is fairly small, but the space available is used with great effect. It took us some time to learn how to use the wood-fired cooker and we still sometimes forget to check the battery store before using electricity. The garden is very easy to maintain and I don't think I have ever eaten so well, or felt so fit.

Life in the village starts, at least for us, at seven o'clock. The children make sure everyone else is awake by collecting the day's eggs from the chicken hutch close to the back of the house. By nine o'clock the children are in school and it is quiet enough to digest the news in the Erg papers. Two days a week Jean helps at the school. Most wives, and quite a few husbands, help at the school at some time, so although there are only three full-time teachers there are usually a dozen or so adults at the school to look after the sixty or so children. At least one day a week we go to Joule market with Galen. We made our first visit to try to buy the things we thought we needed. Now we go to enjoy the atmosphere and watch the bargaining.

Only a small percentage of the transactions on the island involve the exchange of cash. Most are barter exchanges performed according to a fairly rigid formula. The first stages in any deal involve denigrating the product offered by the other person. Once the contestants' vocabularies are exhausted the serious business begins. Normally this involves establishing the government list price of each article, settling any differences with other goods or occasionally with cash. The two main government shops, one selling food, the other tools, machinery and equipment, are adjacent to the market square. It is in these shops that money changes hands and where most islanders buy the fish, meat and cheese produced by the government farms.

We are still battling to understand the prices of things on the island. Some things, for instance cigarettes and books, are much the same price as anywhere else. Clothes, shoes and some food

# A Parable: The Isle of Erg

items are very much cheaper, but tools, machines and any type of fuel are incredibly expensive. We sat down the other evening to try to work out how much everything did cost. We concluded that if we tried to live the way we used to in London we would need an income four times larger than our xat income. But we haven't noticed a drop in our standard of living; just a change.

Our xat income is exactly 1,000 kwats a week, 300 each for Jean and me and 200 for each of the children. We drew up a table of all the things we spent money on last week. It contrasts sharply with our London bills:

|  | Erg (kwats) | London (£s) |
|---|---|---|
| Rent and rates | 200 | 17·00 |
| Housekeeping: |  |  |
|   food | 200 | 10·00 |
|   cigarettes | 20 | 2·30 |
|   drink | 40 | 3·70 |
|   general | 50 | 7·90 |
| Fuel (coal/ electricity, etc.) | 200 | 2·30 |
| Books, post, telephone | 80 | 1·60 |
| Car tax, petrol, etc. | — | 5·00 |
| Other travel | 10 | 1·60 |
| Clothes, shoes, etc. | 100 | 6·20 |
| Cinema, meals out, etc. | — | 2·80 |
| **TOTALS** | 900 kwats | £60·40 |

As you can see we have managed to stay within our income, but this is only because more than half our food comes from the garden and it is the middle of summer, so our fuel bill is smaller than normal. If we had bought all our food in a government shop it would have cost between 600 and 700 kwats, and if the fuel bills had been anything like those we had in London we would have needed 1,600 kwats to pay them. With electricity costing 4 kwats a unit and coal 400 kwats a hundredweight it is not surprising that all the Erg houses have windmills connected to electricity generators, solar roofs for water heating and wood-fired cookers.

The other big expenditure item in our weekly budget is the rates. I went down to the rating office in Brittu to find out how the system works. I spent a long time with Vardamir, the chief rating officer, talking about the Erg system. He explained that the rateable value of a property was decided on the basis of how many 'resources' it used. If nothing was done with the plot of land on which our house stood we would have to pay the maximum rates, which were 400 kwats/week. This was reduced by 100 kwats because of the house, by a further 100 kwats because the garden was intensively cultivated and by 50 kwats each for the solar roof and windmill. I'm still not clear whether the rates could have been reduced to zero by other means, but I did learn that further rate reductions could be claimed if a goat or a cow was kept on the property or if the rainwater was collected and purified.

This scheme gives a property owner a double incentive to exploit the property. In the first place he reaps the benefits directly, in the form of food or fuel; and secondly his rates are reduced. Vardamir explained that the important feature of the scheme was that if you did not produce things for yourself then the government obtained an extra income which it could use to produce the things which you needed but had not produced. In theory this means that if everyone stopped exploiting their own property the extra government income would be sufficient to produce enough food and fuel to keep everyone alive. In practice this would also need a substantial increase in the conscription period that was compulsory for everyone. However, according to Vardamir exactly the opposite is happening. Everyone is finding new ways to produce more of the things that they needed, so that the government farms and industries need to produce less and less. The Erg islanders seem to have the phrase 'every man an island' to heart.

3

*This final report from the Isle of Erg describes the operation of the major industries and the role of private enterprise in the Erg economy.*

The total isolation of the island in 1930 had an enormous impact on its industries. The islanders had already had a taste of the problems in 1922 when, by their own choice, they were without fuels and materials for almost two years. But they did not learn from their experience. As soon as quasi-normal trading relationships were established with France the island returned to a mood of optimism. Their agricultural system was quickly mechanized and by the time Wall Street crashed they could boast one of the world's most advanced agricultural systems – and probably the world's smallest stocks of oil. The small steelworks alongside the iron-ore mines were expanded when the ore shipments to Italy were stopped, but the industry could not provide all the island's needs. Similarly the coal mine, opened in 1870, was improved after 1922, but by 1930, when the islanders realized it was their only source of fuel, the mine could provide only about a quarter of their needs.

These three industries, agriculture, coal and steel, are now the backbone of the Erg economy. They are modern industries, even by European standards, owned by the government and manned by full-time workers on industrial service. There are a number of light manufacturing industries, most of which are also government-owned, producing agricultural machinery, chemicals, glass, paper and some household goods. Most of the household goods used by the islanders are produced by the thousands of craftsmen working for themselves. It is this combination of government-owned heavy industry and individually-owned manufacturing enterprises which restricts the opportunities open to businessmen like Malvegil and Morgoth. With a very stable consumer demand there is virtually no way for businesses to accrue the capital needed for growth.

In most industrial countries the most important growth sectors of the economy are those associated with motor cars, TVs, hi-fi equipment and other consumer durables. There are no markets for these commodities on the Isle of Erg, mostly because they all consume fuel, which is very expensive here. To give one example, a colour television probably uses between one and two units of electricity each evening. Over a week this would run up a bill of 50 kwats, which is as much as we spend each week on cigarettes and wine; in England this is equivalent to £4–£5 per week. If anyone tried to run a motor car on the island they would be faced with petrol at 50 kwats a gallon. All this would be changed if Malvegil can persuade the islanders to open the doors to world trade, particularly the import of oil.

The Minister for Industry, Kaycal, sees this as a quick route to disaster. Over the last twenty years he has been the principal architect of the Erg economy, which produces all the things needed by the islanders from a mere 150 thousand tons of coal, which is a third of the daily coal consumption in the United Kingdom. Even so the coal mines require almost half of the two thousand or so men on 'industrial service'. The most difficult problems in all the government-run industries are associated with the steady decrease in the length of industrial service. In the 1930s the conscription period was ten years, which meant that at least a quarter of the working population were on service at any time. Now the conscription time is down to four years. With the new technologies being used in the major industries this has caused a few training headaches; it takes almost a year to train a skilled machine operator. There has been some increase in the full-time government employees, so that now not only the plant managers but also some of the foremen stay in their job for most of their working life. This problem is going to become more serious as time goes on because recently fewer and fewer children have been choosing to stay on at school and qualify for a permanent position in industry.

There has been no equivalent drop in the number of applicants for the teaching and medical professions. The island is very well endowed with hospitals and schools and has twice the number of doctors per 1,000 population found in the U.K. Although there are also more teachers per 1,000 population, the education system runs into local difficulties because it has to operate on small catchment areas. More women are now taking up part-time teaching in these local schools and the government has recently set up a large retraining scheme to make this more effective. If this does succeed then it will also release some of the full-time teachers to take up positions in the special colleges where the future doctors, teachers, civil servants and industrial managers receive fairly intensive training over a period of five years.

When I spoke to Kaycal about the operation of the Erg economy he stressed the importance of the agricultural system. Amongst a mass of statistics which he used to explain the system the most surprising was that almost half the food needed by the island's population was grown in the 30,000 acres of private gardens. Kaycal also stressed the importance of keeping the Erg population close to today's total. 'At the moment our population density is only a quarter of a typical European country. This makes it easy for us to grow more than enough food without resorting to intensive methods of agriculture. Our government farms have a total area of 150,000 acres and could easily feed the entire population.' The fact that these farms have to produce only half the food needed allows them to produce things other than food. The farm I visited, close to Miltoe in the north of the island, produced a substantial surplus of alcohol from the fermentation of straw and sugar beet grown for the purpose. Much of the alcohol was used as a fuel by the impressive array of farm machinery. The surplus was sent to a new chemical plant close to the farm, which produced medical products. Other farms use their straw for making paper or a constructional material similar to chipboard. Most farms have enough spare land to be able to keep a few horses, which are used to transport the produce into Joule and the other towns of the island.

The full-time government employees, like Gilraen, who managed the Miltoe farm, are given a special home without the normal one- or two-acre garden. These 'job houses' do not have the same high rates as a normal house. While men and women are on industrial service they are given an extra xat allowance to enable them to pay their higher rates because they cannot keep their gardens cultivated while working. Workers on industrial service can choose to move to job houses since there are always a number of them on every farm and close to all the important industries. Small towns and villages also have job houses for the teachers, doctors and administrative officials.

Yet another disincentive to the setting-up of private businesses on anything but an individual level is that private firms are not allowed to put up job houses. This is why Malvegil is campaigning so vigorously against the housing regulations. Well-established private firms, such as Isoar Publications, have their workers living close to their factories. This particular firm also run one of the few transport schemes on the island and uses its delivery vehicles to collect some of its workers from further afield.

At the root of the business community's problems is a phenomenon almost unknown in industrial countries, contentment. The people on this island are happy with their way of life. They only have to work very hard for a four-year period and for the rest of their life they can practise a craft of their own, choosing as much or as little as they want. In our village we have a rabbit breeder, a plumber, two carpenters, one of whom specializes in extraordinarily beautiful furniture, a cycle repairman, two windmill workshops, Galen, who provides a local transport service, a sort of blacksmith-come-tinker who repairs and sharpens all kinds of tools, and the normal potters, weavers, etc. All these activities are based on the satisfactions of serving the community and making something well. Against this background it is easy to see why Malvegil and Morgoth feel frustrated. They cannot persuade the islanders to want a different way of life. From an islander's point of view a leap into the twentieth century looks like a move from a garden of plenty into a quagmire of problems.

# 2 Energy Currency

Not possessing the wit to continue with the parable of the Isle of Erg, I must now come clean and explain what this book is about. Its major aim is to explore the energy options available to the U.K. This is a much wider goal than encompassed within the debates started by the 'energy crisis' in 1973. Most of these debates are concerned with where future energy supplies will come from, how much they will cost and whether or not they will affect the balance of payments. Sometimes the debates are extended to include issues such as the safety of nuclear reactors and whether windmills or schemes for harnessing tidal power are realistic. But only a few people are seriously questioning or trying to evaluate exactly how much energy we *need*. Our long history of low prices for energy led us into a situation where we used it almost without thinking. Now that supplies are either expensive or threatened – or both – our immediate reaction has been to look for alternative sources of supply.

I want to present a different type of approach to energy policy which starts by looking very carefully at energy demand. Exactly how much energy do you *need* each day? Could you reduce your needs by improving the insulation of your house or buying a smaller motor car? How much energy is needed to produce all the things such as cars, food, clothes – even books – that you buy? To answer these questions we will have to spend some time exploring the subject of energy analysis, which provides a method of evaluating these quantities. Once we know how much energy is used in making a loaf of bread or a motor car, then we can base our estimates of energy needs on our demands for cars and bread.

## 22  Fuel's Paradise

There are two other subjects which will also figure strongly in our discussions of energy policy options. The first is associated with technical efficiency and the scientific laws describing the conversion of energy. This is an important topic, since there is no doubt that technical improvements can reduce our demands for all types of energy. However, unless we also understand the basic science underlying the technologies of energy use and conversion we could make some silly mistakes. This is also true if we ignore the final subject which enters the debate, namely the relationship between energy use and climate. This is only one aspect of the much wider subject of the environmental effects of energy use, but it is of particular importance when looking into the future of energy supply and demand.

When we have extracted all the pertinent information from these three subjects – energy analysis, the science of energy use and its climatic effects – we will be in a strong position to examine the energy policy options available to the United Kingdom. We shall still have to consider sources of supply and factors such as the balance of payments, but these will be cast in a much broader framework.

The description of the Erg community and social system in Chapter 1 was deliberately utopian, perhaps even extremist. I felt this necessary in order to counteract the strong fears which are associated with the notion of reducing total energy consumption. The events of late 1973 have had the salutary effect of increasing the general level of awareness of the importance of energy. Petrol shortages in most industrialized countries and the three-day working week in the U.K. were felt as uncomfortable effects of an interruption in fuel supplies. This was exaggerated in the history of Erg to the point where the scarcity of energy dominated the operation of the economy. The result was a totally 'energy-conscious' society in which every effort was made to conserve energy and exploit the energy sources available. This was also reflected by the thinly disguised 'energy currency' used. A 'kwat' is equal to one kilowatt-hour of energy and the 'cost' of commodities in the Erg economy was made equal to their total energy cost.

The concept of 'energy cost' is important in considering how energy is used and whether you or I need more or less energy.

Briefly the energy cost of a commodity equals the total energy required for its production, including the energy used in supplying materials, building factories and operating machines. But all this is taking for granted the concept of energy. Before getting involved in the ideas of energy cost and energy consciousness let me clear away some of the misconceptions which are rampant in the popular reporting of 'energy problems'.

Energy is a word with many different meanings. Active individuals are said to have 'a lot of energy'; when the supply of oil is reduced we are told there is an 'energy crisis' and individuals are urged to 'conserve energy'. Perhaps the confusion is best illustrated by noting that several eminent scientists have urged government to pass legislation 'to conserve energy' while at the same time they teach students of physics and chemistry that one of the laws of nature is that energy is always conserved. Clearly their exhortations in the lecture theatre have a different meaning from those addressed to politicians. To make the distinction clear it is necessary to establish what is meant by the scientific use of the term and later to examine the socio-technical meaning.

Energy is not something that can be experienced by touch, sight, sound or smell. It is an idea. The idea, or mental construct, of energy is used by scientists to explain certain types of events. To appreciate its explanatory power I would like to digress and consider three situations in which a car is made to move along a road.

The first situation is one you may have experienced if you possess a car whose battery has gone flat. It is the situation in which you have to push the car along the road. The significance of this for our present purposes is that you know it requires considerable effort to push the car along the road. It's hard work. Not surprisingly, therefore, if you subsequently see a car travelling along the road you assume that something is replacing your efforts.

The second situation is a hypothetical one in which the car is made to travel along a short stretch of road by connecting it to a falling weight. Let us assume that you have constructed a Heath-Robinsonish device whereby the car has fastened to it a length of rope which passes over a series of pulleys to the top of a suitable cliff. Now let's assume that you collect, from the bottom of the cliff, a massive rock. You have to do a lot of hard work carrying this rock to the top of the cliff. Once there you tie the rope to the rock and push the rock over the cliff. If you collect a suitably massive specimen, the falling rock will pull the car along the road.

The significance of this example is that it was once again *your* efforts which made the car move, but this time you didn't touch the car. Instead you carried a heavy lump of rock to the top of a cliff. Clearly the efforts you invested in lifting the rock were somehow later transferred to the car and it was caused to move. But as you also know the rock was not any different at the top of the cliff from what it was at the bottom. It was not heavier, nor larger, nor denser. Yet somehow it stored your efforts and later released them and made a car move along the road. This series of events is 'explained' by the use of the idea of energy as follows:

When you performed work in raising the weight to the top of the cliff you increased the *gravitational energy* of the weight. This gravitational energy was stored by the weight being in a higher position than initially. When the weight fell over the cliff its gravitational energy decreased until, at the bottom of the cliff, it returned to its initial value. The gravitational energy given up by the weight was tranferred to the car, which is then said to possess *kinetic energy* or energy of motion.

So it is the stored work (done by you) which is called energy and the subsequent sequence of events is described by a series of energy transfers. The usefulness of this way of describing events will, I hope, become clearer as we proceed. The final situation in our discussion illustrates one of the benefits.

Obviously the most common way of making a car move along a road is to use the engine under the bonnet. This is also the socially significant mode of propulsion. Clearly the car would never have caught on if the only way of making one move was to lift a lot of rocks up a cliff! But what is it that replaces your efforts when the car is being driven by the engine? Obviously it is the petrol. In some way the combination of petrol and engine is able to replace all the pushing and shoving you have to do to make the car move. The petrol is said to contain *chemical energy* which can be released by combustion. Using the mechanical apparatus called the engine this chemical energy can replace either the gravitational energy of a rock at the top of a cliff, or your human energy, in making the car move. Thus the idea of energy emphasizes the equivalence of apparently very different things. It says that a gallon of petrol, a rock on the top of a cliff and a strong man are all equivalent in that they can perform a task of work, namely make a car move. It is precisely because of this ability to express equivalence that the idea of energy is used in every branch of science and technology – all that changes is the particular manifestation.

Later we will have to look at other aspects of the scientific concept of energy so as to establish the rules for transforming one type of energy to another and so on. But now it is important to distinguish the scientific idea from the socio-technical use of the word. It is scientifically accurate to state that all processes, all forms of motion, all work, require the conversion of energy from one form to another. Thus it requires energy to shape a piece of metal, to propel a vehicle along a road, to increase the temperature of a room and so on. Not so obvious is the fact that an energy input is required to allow plants or animals to grow and to convey information from one place to another. Energy is thus a universal requirement and is significant in all aspects of our lives. In industrialized societies the productive abilities of men have been increased by the use of particularly useful non-human energy sources, namely coal, oil and gas. There are many other types of energy, in the scientific sense, all around us, things like sunshine, winds and hot air. Although sunshine and hot air are sources of energy they are not technically useful in the sense that our technology cannot harness them to do work. The problems which arose in industrial countries in 1973/4 were due not to a shortage of energy but to a shortage of *technically useful energy*.

For the sake of clarity I will refer to 'technically useful energy' from now on as *fuel*. Thus our present society is dependent on a supply of *fuel* and it is *fuel* shortages which cause economic chaos. Again we must delay further discussion of the idea until later, since the purpose of this chapter is to point out the general aims of the book.

Earlier I pointed out that the Erg currency system was based on the energy cost of commodities. With the above distinctions in mind this is better described as the *fuel cost* of commodities. The

26 *Fuel's Paradise*

idea of fuel cost is central to the considerations in later chapters, so I want to explain what it means now.

The first step is to examine the factors which contribute to the financial cost of commodities. The financial costs of producing something, a motor car say, are divided between payments for labour, for materials, for transport, for fuels and for capital equipment. The payments for 'capital' can be divided into three types, namely the purchase of actual machinery or plant, the purchase of land and payments to 'profits and interest charges'. This complete set of 'factor inputs' is shown in the bar chart in Figure 1.

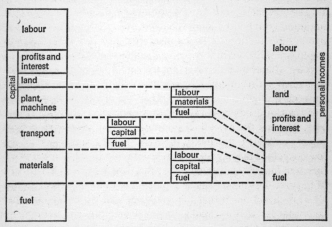

Figure 1  The division of the factor inputs into four basic inputs.

Also shown in Figure 1 are the subsidiary bar charts which divide the payments for transport, materials etc. into their 'factor inputs'. If this process of subdivision is continued then ultimately the total cost can be divided into the four basic inputs, 'labour', 'land', 'profits and interest' and 'fuel'. The aim of an energy analysis is to evaluate the total energy content of the purchased fuels shown in the right-hand side of Figure 1.

Energy analysis does not attribute any 'fuel cost' to the other basic inputs, which are land, labour and profits. This does not mean that these inputs are not important; what it means is that these inputs do not involve any fuel implications. This is fairly obvious for land, but is less obvious for 'labour' and 'profits and interest'. In both these cases the financial payments are made directly to people, in the first case in return for work, in the second case in return for the loan of funds. These payments therefore represent personal incomes, and since these incomes are subsequently used to purchase fuels it might be thought that they should be given a 'fuel cost'. There are three reasons why this would give misleading answers.

The first reason is that in our society individuals are given sufficient income to feed, clothe and keep themselves warm whether or not they are employed. Thus if you reduce the number of people employed (and hence reduce the payments to labour) you will not significantly reduce the fuel consumption of the U.K. The unemployed will continue to consume fuel both directly and indirectly. The second reason is that the extra fuel (or the extra food) consumed by an employed person as compared to an unemployed person is negligible compared to other fuel inputs. At best, energy analysis can produce answers with an accuracy of about 10 per cent, whereas the extra fuel consumed by an employed person would be about 1 per cent of the total. Finally, if any part of the fuel consumed by people were counted as an input to the process then that quantity of fuel would be double-counted in the analysis of the overall system. (This is explained in more detail in Chapter 4.) This would mean that the results of such an energy analysis could not be used to predict energy demand, which is the main reason for including the topic in this book. So for our purposes it would be quite wrong to attribute any fuel cost to personal incomes.

To illustrate the results of a typical energy analysis, Table 1 gives an approximate breakdown of the factor inputs needed to produce a 1,500-cc car in 1968. The total cost, £400, represents the manufacturer's value and does not include retail costs or taxes. The £72 paid to the steel industry purchased just over three quarters of a ton of steel in the form of rods, sheet, wire etc. A similar breakdown of the steel industry shows that 16 per cent of its costs are due to payments to the fuel industries. Thus of this £72 about £11·50 goes towards the purchase of fuels. Similarly almost

20 per cent of the money paid to the glass industry was subsequently paid to the fuel industries. Following through all these purchases of fuel we find that the total money paid to fuel is about £34, which is about 9 per cent of the total value. In physical terms it is possible, knowing the energy content of the different fuels used, to work out the total energy content of these purchases. They are shown in kilowatthours (kWh) in Table 1 and add up to 26,100 kWh. On the Isle of Erg this car would have cost 26,100 kwats!

*Table 1   Factor costs of a 1,500-cc motor car (1968)*

|  | £ cost | Energy content (kWh) |
|---|---|---|
| Labour | 120 | — |
| Capital |  |  |
| profits and interest | 76 | — |
| plant and machines | 22 | 1,120 |
| Transport | 7 | 350 |
| Materials |  |  |
| iron and steel | 72 | 8,640 |
| glass | 4 | 590 |
| other materials, etc. | 91 | 9,170 |
| Fuel |  |  |
| oil, coal, gas | 4 | 3,450 |
| electricity | 4 | 2,780 |
| TOTAL | £400 | 26,100 kWh |

Although this description of evaluating fuel costs has accurately portrayed the general principles, it has glossed over a number of serious difficulties. For example the production of steel involves the use of coal, so to find the fuel cost of steel I first need to know the fuel cost of coal. However, the production of coal involves the use of steel as pit-props and machine parts, so to find the fuel cost of coal I first need to know the fuel cost of steel. In fact this only complicates the situation, it does not make the problem insoluble. It means that 'fuel accounting' turns out to be slightly more elaborate than monetary accounting.

The fact that this method of energy analysis has not included any contribution from 'labour', 'land' and 'interest' is consistent with our aim of understanding how fuels are used. However, it makes nonsense of any idea of a currency based on 'fuel costs'. The use of 'fuel cost' as the basis of a currency is only even plausible on the Isle of Erg because there fuel is a very scarce resource. Furthermore in this fictional example the population is assumed to value the security of self-sufficiency higher than the value of increased material consumption. Neither of these assumptions holds for any real society of which I am aware. In fact the relative financial cost of different resources, basically land, manpower and fuel, in a modern society is supposed to reflect their relative scarcity. Only in a society where land, funds and manpower are in abundance would 'fuel costs' actually be a rational basis for determining the relative values of commodities.

In practice no economic system works in the ways described in economic textbooks. Money and commodities have more effect than simply allocating scarce resources or determining the relative values of items. The accumulation of capital is associated with a degree of power over other men. The formation of workers' unions means that wage rates do not reflect the scarcity of manpower but rather the bargaining strength of the union. The monopolistic control of particular mineral resources may cause them to have a higher price than in textbook theory, and of course military power can force workers or monopolists to relinquish some or all of their economic power. All these effects are largely ignored in economic theories, and were ignored in the utopian description of an energy-conscious society in Chapter 1. I do not think it is possible to devise a theory of resource allocation which will either overcome or take into account all these 'non-economic' factors.

The use of the idea of 'fuel cost' in the following chapters is *not* an attempt to suggest either a more rational or a more humane way of allocating resources. Such a system may arise sometime in the future when for some reason fuel is the most restricted resource. The purpose of introducing fuel cost into our discussions is to evaluate the ways in which industrial societies actually use fuel. It allows me to construct different types of future society and to indicate which require more and which require less fuel supplies. Only in this way can we begin to answer the important questions, namely, 'Do we *need* more fuel supplies?' and 'If so, who needs them and for what purpose?'

# 3 Why We Need Fuels

John Maynard Keynes divided man's requirements into primary needs, defined as those of which you would be aware without any knowledge of other men, and secondary needs, which arise because of your knowledge of other men. Primary needs are those related to the chemistry and biology of man and consist of water, food, warmth and shelter. Secondary needs include psychological requirements such as love and companionship as well as cultural requirements related to status and communication. Without the satisfaction of our secondary needs life would be uncomfortable, perhaps intolerable; without the satisfaction of primary needs you could not live. Most fuel used in the world today is used in satisfying secondary needs, though for most of mankind fuels also play a role in satisfying primary needs. Although most of this book will be concerned with the profligate use of fuels in industrial societies I want to start this chapter by considering the more mundane uses of fuels in non-industrial societies.

To the best of my knowledge the only group of men who live without fuels are Eskimos, and this is possible only because they show two remarkable adaptations. Firstly, they rely entirely on clothing and shelter to maintain their body temperature; secondly, they eat raw food. These adaptations are essential to survival in an Arctic climate, since for most of the year there are no readily available fuels. The Eskimos' ability to maintain bodily warmth using only furs and shelter from low temperatures and cold winds serves to remind us that humans are adaptable and that feelings of discomfort are relative. However, their adaptation to eating raw food is more special. The extreme cold of the Arctic preserves food in a wholesome state. In temperate or tropical climates it is essential to cook food, especially meat and fish, since otherwise it would rapidly deteriorate owing to bacterial, fungal or insect infection. Until recently it was also essential to boil drinking water for similar reasons. So, Eskimos apart, fuels are essential for man's survival in that they provide warmth and enable him to cook food, thereby making it more wholesome and in some cases more digestible.

Fuels, or other non-human sources of energy, are an essential ingredient in what has come to be recognized as civilization for another reason. This is because man can greatly increase his productivity by harnessing useful energy. The influence of non-human energy is most marked in the most basic of human activities, namely the production or collection of food. Without the use of an external source of useful energy, either in the form of fuels or draught animals, man's ability to produce food from a given area of land is limited by his muscle power. If the population density on the land is low enough this is not a serious constraint. Sahlins (Sahlins, 1972) has shown that in terms of the amount of effort needed to produce a 'comfortable existence' stone-age society was an 'affluent society'. However, as the population density increases, more food has to be produced from the same area of land. This pressure on food resources caused a number of technical developments (Wilkinson, 1973) all aimed at increasing man's productivity. The most significant developments have been those which have added to man's muscle power so that more work could be done in the same time. These additions, initially in the form of animals, now in the form of tractor fuels, have enabled man to continue to feed himself and still have time for leisure activities (such as reading and writing books!). Ultimately we are dependent on our highly mechanized and fuel activities for our leisure time. Thus in a primitive agricultural community the food output per acre is very much lower than in an industrialized agriculture. One primitive farmer produces enough food for three to seven people, depending on local conditions. One industrial farmer can provide a better food supply for about a hundred people. Even if we take into account the four or five people employed in the industrial system in providing the farmer with tractors and distributing his output, the industrial system achieves about twice the output of the primitive system (Leach, 1975). This is an extremely important factor, since it allows a much larger proportion of the population to produce other artefacts and spend time on non-essential activities.

The role of fuel and other energy sources in agriculture can best be illustrated by examining the ratio of energy output to energy input – the energy ratio – for different food-collection systems. Men, like all other animals, require food to provide them with energy. The chemical energy contained in food is used in the body to maintain essential functions, such as blood circulation, to provide energy for muscular movement, to provide energy for the synthesis of new tissue and to keep the body warm. The art of being a successful animal is to be able to collect more energy in food than has to be expended in collecting the food. The approximate measurements which have been made indicate that most animals achieve an energy ratio of between 7 and 25 (Lawton, 1973). In primitive agricultural systems the ratio of food energy gathered to human energy expended varies between 10 and 30, with most crops falling in the range 15 to 20 (Leach, 1974). These figures indicate that most 'primitive' farmers do not need to work very hard to provide their basic needs, but they do use considerable areas of land.

In contrast to these more or less 'natural' food systems, a modern intensive farm may produce 1,000 units of food energy for every unit of human energy expended (Pimentel, 1973). Not only is the energy return much higher but so too is the yield per acre. Over the past twenty-five years the crop yield per acre has doubled for most of the crops harvested in the U.K. and U.S.A. These advances have been possible only because of an enormous non-human energy input to the food-production system. Until about a hundred years ago the only non-human energy input to agriculture was provided by draught animals, principally horses. Using horses and an intensive crop-rotation system the ratio of food energy produced to human-energy input was between 100 and 200. The increase from this figure to the present 1,000 is due to the use of tractors, which burn diesel fuel, and fertilizers, insecticides and herbicides, which are products of the petro-chemical industries. If the fuel energy is counted as an energy input to agriculture, then the energy ratio changes from a profit of 1,000:1 to an energy loss of about 1:2. This has led one ecologist to remark that man no longer eats potatoes made from solar energy; now he eats potatoes partly made of oil (Odum, 1971). The total energy ratio, including the fuel inputs, for a wide range of crops is documented in Table 2. Also shown are some rough estimates (Steinhart, 1974) of the energy ratio of the entire U.S. food system. Figure 2 shows the crop yields, manpower input (on the farm) and total energy input for the U.K. since 1940. The costs (in fuel terms) and benefits (in increased productivities per acre and per man) are obvious.

Table 2  Energy ratios for various crops

| Crop | Energy ratio | Reference |
| --- | --- | --- |
| Potatoes (U.K.) | 1·0–1·2 | (Leach, 1974) |
| Sugar beet (U.K.) | 0·5 | ,, |
| Wheat (U.K.) | 2·2 | ,, |
| Barley (U.K.) | 2·0 | ,, |
| Corn (U.S.A.) | 2·0 | (Steinhart, 1974) |
| Soyabeans (U.S.A.) | 2·2 | ,, |
| Corn (U.S.A.) | 2·82 | (Pimentel, 1974) |
| Rice (Indonesia) | 45·0 | (Steinhart, 1974) |
| Rice (China) | 40·0 | ,, |
| Rice (Burma) | 25·0 | ,, |
| Rice (China) | 53·0 | (ref. in Leach, 1974) |
| Taro-Yam (New Guinea) | 20·0 | ,, |

Before the advent of tractors and cheap fuel the major source of energy on the farm was provided by animals, especially horses. However, animals have to be fed and land has to be given over to providing them with food. A large part of the increase in total food production over the past fifty years has been due to the replacement of horses by tractors, thereby making many more acres available for human food production. With the continued growth in human populations this transition provided an essential boost to total food output and the transition could be reversed only if populations decreased close to the levels of fifty years ago. So although the use of fuels in agriculture is not essential in that crops do not need fuel to make them grow, it is essential in maintaining food supply for the now much larger human population.

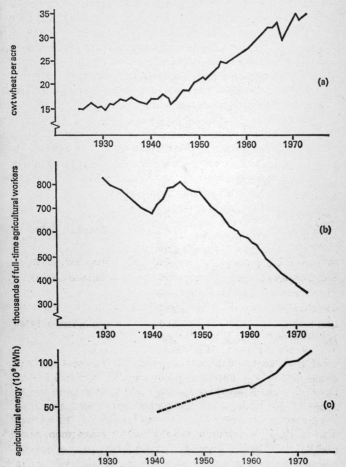

Figure 2 (a). UK. wheat yield (H.M.S.O., 1966); (b) full-time farm workers (H.M.S.O., 1966) and (c) total energy imput to U.K. farms (Leach, 1975).

Fuel-intensive agriculture is only common practice in the industrialized countries. In non-industrial countries traditional farming methods, based on use of animals and crop rotation, still dominate. It has often been suggested that the world would be able to support a very much larger population than the present 3·5 billion if only these traditional agricultural systems could be replaced by modern industrial systems.

For example, the Professor of Organic Chemistry in the University of Cambridge has stated 'that if all the land *now* cultivated (i.e. without making allowance for the vast tracts of land not yet brought under cultivation in Latin America, Australia and parts of Asia) were to be cultivated as efficiently as it is in the Netherlands, the world could support 60 billion people – i.e. ten times as many as are expected by the end of the century' (Beckermann, 1974). This type of view is based on the observation that in the U.K. the agriculture and food industries account for only 2 per cent of total fuel consumption. However, this ignores the much larger quantities of energy used indirectly in the food and agricultural sector in the form of fertilizers, insecticides, machinery, transport and packaging.

If all these items are included in the energy calculations, then, as will be shown in Chapter 4, about 17·5 per cent of all the U.K.'s fuel is used in the production of food. Even this figure is not complete, since it does not include the energy expended in other countries in producing commodities such as fish-meal and phosphate fertilizers which are important inputs to U.K. agriculture. On this basis, if every field in the world were cultivated to the U.K. average, something like 50 per cent of the present total world fuel consumption would have to be devoted to food production. This is one of many cogent reasons why the whole world cannot go over to a fuel-intensive agricultural system as practised in industrialized countries (for other reasons see Odum, 1971; Leach, 1974; and Slessor, 1973).

Although industrial agricultural practice cannot be extended worldwide, the graphs shown in Figure 2 do illustrate well the increase in productivity which is possible with increased fuel use. In general it is foolish for man to have his productivity constrained by his muscle power, since by using other sources of power his ability to think and process information can lead to much greater productivity. The use of fuels amplifies man's productivity by allowing his information-handling ability to be extended. The role of technology has always been to develop tools, artefacts and techniques for accomplishing tasks more easily and more quickly. In simple tasks, such as digging an area of land for cultivation, the role of tools, technology and fuels are obvious. It would take a very long time to dig over say a quarter of an acre field using only usefully shaped pieces of flint and one's own muscles. The use of a spade, a carefully shaped implement made of wood and metal, reduces the time perhaps ten-fold, to about half a day. Given a horse and plough the same area could be dug over in a couple of hours and with a tractor and multiple plough in less than an hour.

Productivity is always related to time. In our present society time, like anything else, is measured in money terms and productivity is conventionally referred to in terms of the financial cost of labour. The faster something can be produced, the smaller the wage bill per item produced and the greater the productivity. In real terms it is better to consider productivity in terms of useful work accomplished in a given time, thereby avoiding problems of changing wage-rates. It is the real increase in productivity per unit time which enables us to lead a comfortable life and spend time writing and reading books.

Time can also be introduced into the discussions of fuel and energy by using the scientific concept of power. Power is another one of those words with many popular meanings but just one scientific meaning. In science or technology the power of a machine is defined as the amount of energy it converts per unit time. Power is measured in kilowatts (kW) and energy in kilowatt hours (kWh). Thus a machine which converts 30 kWh (kilowatt-hours) of electrical energy into heat energy in one hour has an average power rating of 30 kW (kilowatts). In two hours the machine would convert 60 kWh of energy, in three hours 90 kWh and so on.

The relationship between energy and power can best be explained by analogy with the relationship between distance and speed. *Speed* is defined as the distance travelled per unit time; *power* is defined as the energy converted per unit time. Similarly the total *distance* travelled is the average speed multiplied by the time spent travelling; the total *energy* converted is the average power multiplied by the time of operation.

This analogy can be extended to explain why the power of a machine is usually more important than its energy capacity. The point is that most machines can handle an indefinitely large amount of energy but only at a finite rate. Left operating long enough a 100-watt light bulb could consume all the fuel resources of the world. But that is not the interesting thing about light bulbs. If I purchase a lamp I want to know how much light it gives out, which is related directly to its power rating. In terms of the speed–distance analogy the equivalent point is that motor cars are specified according to their maximum *speed*, not the total distance they can travel (which is in any case indeterminate).

Earlier I pointed out the relationship between increasing fuel use and productivity. In fact there is a more direct parallel between power rating and productivity. If we assume that it takes a fixed amount of energy to make something, say to bend a wire to the shape of a paper clip, then clearly the number of paper clips which can be made in an hour (the productivity) depends upon how much energy can be provided to bending wire in an hour (the power of the machine).

The concept of power can be put into focus by noting the range of power ratings that exist in the world. If I sit quite still my body has to convert food energy at a finite rate simply to keep me warm and alive. Under these conditions my power rating is about 0·02 kW. If I run upstairs as fast as I can, my muscles are working hard and my body has a power output of about 0·1 kW. Some very fit athletes have achieved a peak power output of 0·5 kW, but this can only be sustained for a second or two. Classically a horse has a power output of one 'horse-power', equal to 0·75 kW, but in practice most work horses achieve higher power outputs than this. A simple machine like an electric drill has a power rating of 0·25 kW, and power tools, of the type used in light manufacturing industry, have power ratings between 1 and 10 kW. A motor-car engine has a power rating of about 100 kW, but only a fraction of this is available for propelling the car along the road. Railway locomotives have power ratings of several hundred kilowatts and an aircraft like Concorde has a power rating of 10,000 kW. The very largest machines are those used in electricity generating stations where a single turbine can have a power rating as high as 500,000 kW. This last number is equivalent to the combined power output of five million men working hard! Although such comparisons are not strictly permissible, since there is more to the accomplishment of a task than simply providing the necessary energy, they do emphasize the vast increase in productive capacity which exists in an industrialized nation, with perhaps a hundred large turbines, compared to a non-industrialized country with a population of perhaps a hundred million people.

# 4 How We Use Fuels

In principle all the fuels consumed in an industrial society are used to further the well-being of the people who live in the society. However, most of this fuel is used indirectly. I can illustrate this by calculating the total amount of fuel consumed by an average family in a year. The results, shown in Table 3, are based on the *Family Expenditure Survey for 1968* (H.M.S.O., 1968). The amount of fuel is converted to an equivalent energy using the energy content of fuels. For domestic fuels these are 8,000 kWh/ton coal; 29·3 kWh/therm of gas; 1 kWh per unit of electricity; 44 kWh/gallon petrol and 52 kWh/gallon fuel oil.

Table 3 *Direct personal use of fuels*

|  | (1968, for four-person family) | Energy content kWh |
|---|---|---|
| Electricity | 4,020 units | 4,020 |
| Coal | 2·15 tons | 17,200 |
| Gas | 137 therms | 4,014 |
| Fuel oil | 17 gallons | 884 |
| Petrol | 420 gallons | 18,480 |
|  | TOTAL | 44,598 |

This total of 44,598 kWh/year for the average four-person family corresponds to 11,149 kWh/person/year. The total population of the U.K. is about 55 million, so the total fuel consumed directly by people equals 613,222 million kWh. This can more conveniently be written as $613 \cdot 22 \times 10^9$ kWh.[1] A similar calculation can be done for the total fuel input in the U.K. As shown in Table 4 (below), this was equal to $2,341 \cdot 7 \times 10^9$ kWh in 1968. Thus the direct personal use of fuels accounts for only 26 per cent of total use.

So what happens to the other 74 per cent of the fuels used in the U.K.? The answer is that it is used on our behalf to provide us with goods and services. In the rest of this chapter I want to show how the remaining 74 per cent is used and thus be able to describe the amount of fuel needed for the production of various goods and services.

Essentially what is needed is to be able to work out the fuel cost of all commodities made in the U.K. in such a way as to account for the total fuel expenditure. If this is done successfully, then when we add up all the commodities sold, and multiply the results by the appropriate fuel cost, we should arrive at a total equal to the primary fuel input to the U.K.

To perform this type of analysis we are going to have to get our feet wet puddling around in a sea of numbers and statistics. An important requirement of the statistics used is that they should be comprehensive and give details of the fuel consumption of all industries. At the time of writing the latest detailed statistics available are those published in *Report of the Census of Production 1968* (H.M.S.O., 1971), so all the data will refer to 1968. Although this is now out of date, there are good reasons for believing that the patterns of fuel consumption change slowly. So the picture I am going to paint is probably still valid today, although, of course, the total quantities of fuel have increased substantially since then.

The first steps involve evaluating the total fuel input to the U.K. and finding out who purchases the fuels. The primary fuel input to the U.K. in 1968[2] consisted of 154·6 million tons of coal; 82·9

---

1. To avoid very long strings of zeros large numbers can be represented by writing the significant digits and stating how many places the decimal point has been moved. The number above the 10, in the example in the text 9, indicates the number of places the decimal point has to be moved to the right to get the number in normal form. Thus $2 \cdot 0 \times 10^2$ equals 200·0 and $1 \cdot 34 \times 10^3$ equals 1,340·0.

2. This is data taken from the *Census of Production*. It is slightly different from data for the same year published in the *Annual Digest of Energy Statistics* (H.M.S.O. annual).

## 42 Fuel's Paradise

*Table 4 Primary fuel inputs to the U.K. 1968*

|  |  | kWh | per cent |
|---|---|---|---|
| Coal | 154·6 m. tons | $1,177·3 \times 10^9$ | 50·3 |
| Oil | 82·9 m. tons | $1,029·3 \times 10^9$ | 44·0 |
| Gas | 1,207 m. therms | $35·4 \times 10^9$ | 1·5 |
| Hydro-electricity |  | $3·6 \times 10^9$ | 0·1 |
| Nuclear heat |  | $96·1 \times 10^9$ | 4·1 |
|  | TOTAL | $2,341·7 \times 10^9$ | 100 |

million tons of oil; $30·8 \times 10^9$ kWh of nuclear and hydro-electricity and 1,207 million therms of natural gas.[3] All the other fuels used in the U.K., such as electricity generated in oil-fired power stations, town gas and coke, are known as secondary fuels since they are produced from one of the primary fuels. The primary fuels can be combined into a total fuel energy input using the energy content of each fuel, as shown in Table 4. The energy content of coal is taken as 7,615 kWh/ton, since this represents the U.K. average (the value of 8,000 kWh/ton used in Table 3 was for domestic-grade coal). The nuclear electricity is counted on a heat-equivalent basis for reasons that will be made clear in later chapters. The outputs of the fuel industries are the fuels purchased by you and me or by industry. These are shown in Table 5, together with their total energy content (Chapman, 1974 (a)).

*Table 5 Fuels delivered to consumers*

|  |  | $10^9$ kWh | per cent |
|---|---|---|---|
| Coal | 42·5 m. tons | 323·9 | 19·7 |
| Coke | 9·9 m. tons | 78·1 | 4·7 |
| Oil products | 74·7 m. tons | 918·3 | 55·8 |
| Gas | 51·7 m. cu. ft. | 151·6 | 9·2 |
| Electricity |  | 174·9 | 10·6 |
|  | TOTAL | 1,646·8 | 100 |

3. This is mostly methane obtained from collieries; the remainder is gas from the North Sea.

Comparing Tables 4 and 5 you can see that the mixture of fuels used is quite different from the input mixture. For example electricity represents only 4·2 per cent of the primary fuel input but 10·6 per cent of fuels used by consumers, and coal, which supplies over 50 per cent of the primary input, supplies less than 20 per cent of the fuels used by consumers.

However, more significant is the fact that only 69·8 per cent of the primary fuel energy ever reaches a final consumer. The other 30·2 per cent is used by the fuel industries themselves for operations such as mining coal, refining oil, transporting coal, pumping gas and generating electricity, the largest single consumer being the electricity industry. A detailed analysis of the electricity industry shows that on average it requires 4 kWh of primary fuel input to produce 1 kWh of electricity, an efficiency of 25 per cent. This low efficiency arises for several reasons. There are scientific reasons why you cannot convert fuel to electricity with anything like 100 per cent efficiency. The theoretical limit of conversion efficiency is about 60 per cent and the most modern power stations achieve 35 per cent. The overall average of 25 per cent is due to the fact that there are a lot of old power stations in the U.K. and there are significant losses involved in distributing electricity.

The fact that 4 kWh of primary fuel are necessary to produce 1 kWh of electricity has a lot of important implications which will be discussed later in the book. For the moment its significance is that if you use 1 kWh of electricity in your house you are *in effect* consuming 4 kWh of primary energy. In the terms introduced in Chapter 2 the fuel cost of 1 kWh of electricity is 4 kWh. In order to distinguish fuel costs from energy content I will denote fuel costs by *kilowatt-hours-thermal*, kWht, and energy content by kilowatt-hours (kWh). Thus the fuel cost of electricity is 4 kWht per kilowatt-hour.

In a similar way I can evaluate the fuel costs of all the other fuels dispatched by the fuel industries. For example for every ten tons of coal you consume, 10½ tons of coal have to be mined; the other half ton is used to mine the coal and transport it to you. For every ten gallons of petrol you consume the equivalent of eleven gallons has to come up an oil well – the other gallon being used in transporting the oil to the U.K., refining it, delivering it to your local garage and pumping it into your car. The ratio of energy delivered to final consumers divided by the primary fuel input defines an efficiency for each of the fuel industries. These are shown in Table 6 for three different years (Chapman 1974 (a)).

Table 6  *Efficiencies of the fuel industries (per cent)*

|  | 1963 | 1968 | 1972 |
|---|---|---|---|
| Coal | 95·5 | 96·0 | 95·5 |
| Coke | 75·5 | 84·7 | 88·0 |
| Gas | 64·7 | 71·9 | 81·1 |
| Oil | 80·8 | 88·2 | 89·6 |
| Electricity | 22·0 | 23·9 | 25·2 |

Using these fuel-industry efficiencies it is possible to calculate the fuel costs of all the different fuels used by consumers. For example the fuel cost of one ton of domestic coal is 8,334 kWht compared to its energy content of 8,000 kWh and the fuel cost of petrol is 50 kWht/gallon compared to an energy content of 45 kWh/gallon. Table 7 shows the personal use of fuels evaluated in terms of fuel costs instead of energy content. The new total, equivalent to 15,564 kWht/person/year, represents the proportion of the primary fuel input to the U.K. consumed directly by the 'average person' in 1968. For 55 million such people it corresponds to $856·3 \times 10^9$ kWh/annum, which is 36·6 per cent of the total fuel input.

Table 7  *Direct use of fuels*

|  | (1968, for four-person family) | Fuel cost kWht |
|---|---|---|
| Electricity | 4,020 units | 16,820 |
| Coal | 2·15 tons | 17,918 |
| Gas | 137 therms | 5,582 |
| Fuel oil | 17 gallons | 1,002 |
| Petrol | 420 gallons | 20,952 |
|  | TOTAL | 62,274 |

## How We Use fuels

There are good reasons for using fuel costs rather than energy content for evaluating patterns of fuel use. For instance if you look only at the energy content of consumption you might conclude that not using 1,000 units of electricity would reduce the fuel input to the U.K. by 1,000 kWh. In fact if you don't use 1,000 kWh of electricity the power station does not have to burn 4,000 kWh of fuel, so the fuel input is reduced by 4,000 kWh, not 1,000 kWh. In other words the fuel cost gives you a better estimate of the primary fuel saving you could make by stopping an activity or the extra fuel needed to increase an activity.

By using fuel costs instead of energy content we have now accounted for just over a third of the total use of fuels in the U.K. That still leaves us with two thirds to explain away. This can best be done by dividing the economy into eight fuel-consuming sectors as shown in Figure 3. The input of primary fuel is put equal to 100, so that the numbers on the top of each sector box represent the percentage of total fuel used on an energy-content basis.

The domestic sector refers to fuels purchased directly by households for purposes of cooking, heating, lighting and the operation of domestic equipment. The use of petrol in private cars is shown in the private transport sector and is a fairly tentative estimate since it is not possible to decide exactly how much petrol is used for private motoring and how much for commercial purposes. However, the sum of these two sectors does agree fairly well with our previous estimate of personal fuel use as equal to 26 per cent of total primary fuel.[4] The other transport sector incorporates fuels used for all kinds of freight transport by road, rail, air and sea. It does *not* include transport carried out by an industry's own vehicles. Fuels consumed by, say, the bread industry in their own vehicles is included in the 'food industry' sector. The food and agriculture sector also includes fuels used on farms and in all kinds of food-processing industries. The materials sector, which is the second largest consuming sector, includes all those industries which produce raw materials such as steel, cement, plastics, bricks, paper and rubber. The manufacturing sector includes all those industries which produce commodities, buildings, machines

4. The agreement is not exact since a family with the average income will not consume the average amount of fuel. This is explained further in Chapter 10.

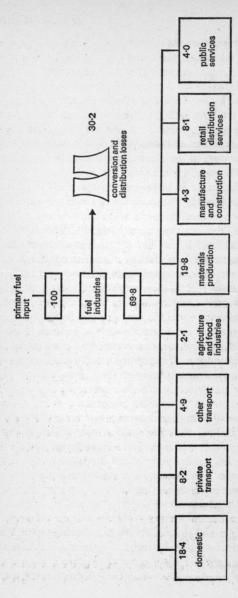

Figure 3 The breakdown of fuel use into sectors according to energy content of the fuels used.

and other types of capital equipment. The retail sector includes the retail and wholesale trades and other service industries such as repair firms, travel agents and barbers' shops. The public-services sector includes all the fuel consumed by the armed services, the police, the civil service, other government departments and local authority activities such as refuse disposal and street lighting.

Now we want to convert this sector division into a 'fuel-cost' diagram so that we can take into account the different mix of fuels supplied to each sector. There is one small problem here which arises because the fuel industries consume some of the output of the manufacturing and materials sectors. For example, oil refineries use chemicals produced in the materials sector and steel equipment produced in the manufacturing sector. This is shown in Figure 4 by the subsidiary input to the fuel industries of 4·2 per cent. This makes the total output of the fuel industries have a 'fuel cost' equal to 104·2 per cent of the primary fuel input. The net output of all the sectors, that is the total output less the quantity supplied to the fuel industries, is equal to 100 per cent, which is what we wanted. (As explained above the total fuel cost of all the final output must equal the total primary fuel input.)

This new division between the sectors, in terms of the fuel cost of the fuels, is considerably more useful for our later analyses than the division in terms of energy content. Notice that our direct use of fuels, in the domestic and passenger-transport sectors, is equal to 38·6 per cent, which is close to our estimate based on the 'average family'. Again the agreement is not exact because the average family does not consume the average quantity of fuel.

Although this breakdown into sectors, which is probably accurate to about 5 per cent, gives a good general picture of where fuels are used, it is not yet in a form in which we can relate fuel use to end products. For example you and I rarely purchase any of the products of the 'materials' sector directly; we haven't much use for steel ingots or flagons of sulphuric acid. The outputs from the materials sector are inputs to other sectors. Steel is supplied to the motor-car industry, fertilizers to agriculture, copper to the electrical trades and so on. In order to appreciate all the fuel implications of an activity, such as producing food, these transactions have to be taken into account. This is done by giving a fuel cost to

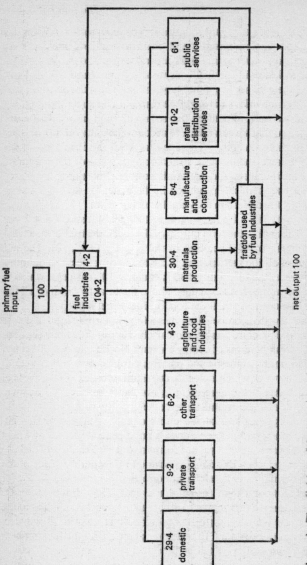

Figure 4 The division of fuel use according to the fuel cost of the fuels used.

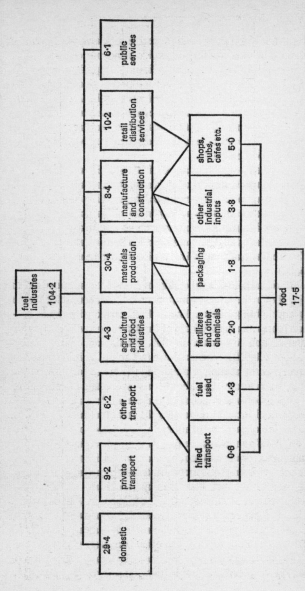

Figure 5 The total fuel cost for providing food, drink and tobacco. There are contributions from five of the fuel-consuming sectors.

the intermediate products, such as steel ingots, and incorporating these in the total fuel cost of end products, like motor cars. Figure 5 summarizes the results of an analysis of this type for the production of food. It shows that in fact food production accounts for about 17·5 per cent of the primary fuel input to the U.K. This total includes the fuels used on the farm, the fuels used in food industries, the fuels used in making tractors, machines, fertilizers and packaging, as well as an estimate of the fuels used in food shops and in transporting food products.

Figure 6 shows a similar analysis of all the fuels used in the provision of passenger transport of all kinds. This incorporates all the direct fuel consumption of cars, buses, ships, planes and passenger trains. Also shown are the fuel expenditures associated with the construction of transport vehicles such as motor cars and aeroplanes. The total of 20·3 per cent represents the total fuel implication of passenger transport. (This is discussed in more detail in the Appendix.)

The use of fuels in the provision of housing is summarized in Figure 7. By far the largest component is the direct use of fuel in heating and lighting houses. The fuel associated with house construction represents the total use of fuels in the provision of building materials and new houses in 1968. For any one house the fuel cost is about 100,000 kWht, as shown in Table 8. This is equivalent to the fuel used in heating and lighting the house for three years. However, since houses last about forty years this investment of fuel represents only a fifteenth of the annual fuel running costs. The 'household fittings' represents the total fuel used in 1968 in the production of carpets, furniture, linoleum and so on. The 'consumer durables' represents the fuel used in the production of electrical appliances such as vacuum cleaners, refrigerators and TV sets. Thus the housing total incorporates all the permanent fixtures in the house and the fuels needed to operate them.

On this type of basis the total fuel use in the U.K. can be broken down into six major categories of activity, as shown in Table 9. These are the provision of food, of housing, of passenger transport, of clothing, the public services and other consumer goods and services. The public services are the armed services, the provision of government and services such as the police force and hospitals.

*Table 8 Fuel cost of a house*

Parker Morris three-bed semi-detached, 100m² floor-space

|  | kWht |
|---|---|
| Bricks 16,000 at 1·6 kWht each | 25,600 |
| Steel 1·2 tons at 13,200 | 15,840 |
| Glass 320 ft² 0·38 tons at 6,277 | 2,385 |
| Concrete 12 yd³ at 630 | 7,560 |
| Cement 2 tons at 2,200 | 4,400 |
| Plaster 3 tons at 900 | 2,700 |
| Timber 4·3 cu.m = 151·9 ft³ at 31·3 | 4,756 |
| Plastics 250 lb. = 0·113 tons at 45,000 | 5,085 |
| Paint 4,700 sq. ft (assume £1 per 100 sq. ft for all coats, thus £47 at 156·2 kWht/£) | 7,341 |
| Copper, etc. 500 lb. = 0.226 tonnes at 15,000 | 3,390 |
| Miscellaneous | 4,000 |
| Total materials | 83,057 |
| Construction energy (deduced from fuel consumption per £ value for construction industry) estimated to be | 19,000 |
| GRAND TOTAL | 102,057 kWht/house |

The final category of 'other goods and services' incorporates such things as cosmetics, books, cameras, records, household chemicals and jewellery, as well as barbers' shops, cinemas, travel agents and banks. The transport sector excludes freight transport, which is incorporated in one of the other sectors. For example the transport of wheat would be included in the 'food' sector and the transport of bricks in the 'domestic' sector.

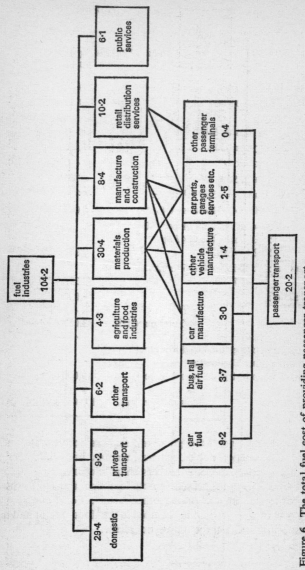

Figure 6 The total fuel cost of providing passenger transport.

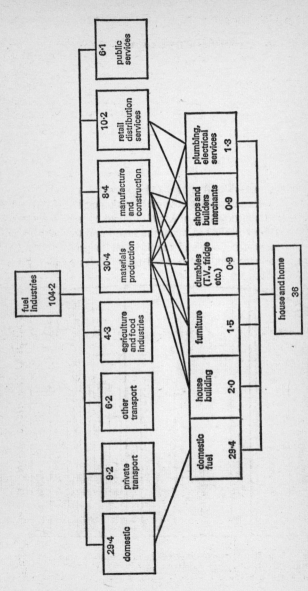

Figure 7 The total fuel cost of 'house and home' including house building, furniture and durables.

## 54  Fuel's Paradise

*Table 9  Fuel use by sectors of final demand*

|  | per cent |
|---|---|
| Provision of food | 17·5 |
| House and home | 36·0 |
| Passenger transport | 20·3 |
| Clothing and footwear | 6·7 |
| Public services | 11·5 |
| Other goods and services | 8·0 |

This type of analysis of fuel use can be carried to a level of much finer detail. For example it is possible to evaluate all the fuel inputs required for the production of a loaf of bread by tracing back the quantity of flour needed to make the bread and the quantity of wheat to make the flour and the fertilizer to grow the wheat. Figure 8 shows the results of such a detailed analysis. Later, when

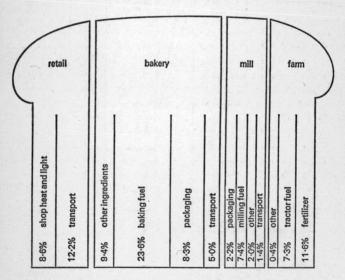

Figure 8  The detailed breakdown of the fuel cost of a standard white loaf. The total fuel cost is 5.6 kWht/loaf.

we want to compare different modes of production or different lifestyles, this detailed analysis will prove invaluable. For example if you envisage an agricultural system operating with organic fertilizers instead of inorganic fertilizers you know that you will save about 7 per cent of the fuel needed to make a loaf of bread, that is about 0·4 kWh/loaf. Not packaging loaves would save 0·5 kWh/loaf and halving the amount of transport needed could save 1 kWh/loaf.

In order that you might get a 'feel' for the relative size of fuel costs, Table 10 gives the fuel costs of a wide range of commodities as produced in the U.K. in 1968. The second column gives the ratio of fuel cost to financial value where the value is the ex-factory price. Thus the value does not include wholesaler's or retailer's mark-up, but does incorporate manufacturer's profits. Both sets of data in this table (shown on the next two pages) will be of use to us later in the book.

## 56 Fuel's Paradise

*Table 10 Fuel cost of commodities (Chapman, 1974 (b))*

| Item | Fuel cost kWht | Fuel cost/value kWht/£ |
|---|---|---|
| Coal per ton | 8,334 | |
| Oil per ton | 13,718 | |
| Petrol per gallon | 49·9 | |
| Gas per cubic foot | 0·41 | |
| Electricity per kWh | 4·2 | |
| House (3-bedroom semi) | 100,000 | 35 |
| Motor car (1,000-cc capacity) | 22,500 | 69·5 |
| Washing machine | 2,100 | 54·4 |
| Refrigerator | 1,450 | 54·4 |
| Vacuum cleaner | 650 | 54·4 |
| TV (colour) | 6,900 | 43·9 |
| TV (black and white) | 1,650 | 43·9 |
| Transistor radio | 430 | 43·9 |
| Record player | 525 | 43·9 |
| 3-piece suite | 5,000 | 50 |
| Dining table | 1,000 | 50 |
| Bed | 820 | 63·5 |
| Loaf of bread (ex-bakery) | 4·4 | 54 |
| Pint of milk (ex-dairy) | 2·40 | 64 |
| Dozen eggs        „ | 5·76 | 47 |
| 1 lb. cheese      „ | 6·2 | 47 |
| 1 lb. bacon       „ | 16·40 | 99 |
| 3 lb. chicken | 36·9 | 100 |
| 3 lb. beef | 30 | 100 |
| 1 lb. fish | 27 | 94 |
| 1 lb. tea | 19 | 74 |
| 1 lb. coffee | 39 | 63 |
| 2 lb. sugar | 7·4 | 123 |
| 1 lb. biscuits | 6·2 | 61 |
| 1 lb. chocolate | 8 | 57 |
| 5 lb. potatoes | 1·6 | 50 |

## How We Use Fuels 57

| Item | Fuel cost kWht | Fuel cost/value kWht/£ |
|---|---|---|
| Can baked beans | 3·1 | 64 |
| 20 cigarettes | 0·4 | 9·0 |
| 1 pint beer | 1·6 | 17·9 |
| 1 pint whisky | 71·5 | 29·8 |
| 3-piece suit | 475 | 45 |
| Dress | 110 | 45 |
| Pair of shoes | 62 | 44 |
| Mackintosh | 145 | 46 |
| Shirt | 54 | 56 |
| Bottle of perfume (150-cc) | 104 | 112 |
| Newspaper | 0·6 | 41 |
| Glossy magazine | 2·6 | 41 |
| Paperback book | 4·5 | 45 |
| Bicycle | 1,700 | 85 |
| Milk bottle | 2·3 | 140 |
| Finished steel (1 ton) | 13,200 | 212 |
| Aluminium (1 ton) | 27,000 | 74 |
| Copper (1 ton) | 12,800 | 32 |
| Cement (1 ton) | 2,200 | 410 |
| Plastic (1 ton) | 45,000 | 213 |
| Glass for window (per sq. ft) | 8·6 | 140 |
| Wooden door | 1·7 | 66 |
| Brick (common) | 1·6 each | 173 |
| Excavating machine | 340,000 | 69 |
| Crane | 320,000 | 45 |
| Ship (100,000 Dwt tons) | $500 \times 10^6$ | 25 |
| Aeroplane (Jumbo Jet) | $20 \times 10^6$ | |
| Locomotive | $1·4 \times 10^6$ | 178 |
| Coal fired power station | $2,700 \times 10^6$ | |
| Nuclear power station | $10,200 \times 10^6$ | |
| Uranium (per ton 0·3 per cent ore) | 383,000 | |
| Oil rig | $2,500 \times 10^6$ | 50 |

# 5  Where It Goes To

The previous chapters have set out in broad terms why we need fuels and how they are used in an industrial nation. Now we have to return to some topics on the nature of energy, in particular the usefulness of different fuels. For example, why do we apparently waste large quantities of coal energy by converting it to electricity, which has an efficiency of 25 per cent? Why is the efficiency of generating electricity restricted to a theoretical maximum of 60 per cent? What are the factors which influence the efficiency with which fuels can be converted to useful work? If we can answer these questions, we shall be in a position both to look at the climatic effects of fuel consumption and to be able to impose sensible limits on future technology. These are both important considerations for the future of fuel use.

There are basically two rules which describe the conversion of fuels to work or other forms of energy. They are called the first and second laws of thermodynamics. The first law is better known as the law of conservation of energy and states that energy cannot be created or destroyed, only converted from one form to another. The second law states that it is impossible to construct a machine that will produce no other effect than the transfer of heat from a cooler to a hotter body. These rather abstract statements are not obviously relevant to our concern, so let me start by showing how they operate in a practical situation.

I can illustrate both laws in action by considering all the energy transformations which take place when a motor car is driven along a road. We have already established that the source of energy in this case is the chemical energy contained in petrol. When the petrol is burnt in the engine this chemical energy is converted to heat. The heat causes the gases in the chamber to expand and in doing so the gas pushes a piston down. When the piston has been forced to the end of its travel the hot gases left in the chamber are expelled down the exhaust pipe. The downward movement of the piston is transmitted to the wheels of the car by a number of mechanisms such as the crankshaft and gearbox. As the car goes faster and faster its kinetic energy, or energy of motion, increases and it is this increase in kinetic energy that represents the useful result of burning the petrol. However, not all the energy in the petrol is converted to kinetic energy of the car. Almost three quarters of it is thrown out with the hot gases ejected down the exhaust pipe. Some of the energy goes to overcoming friction in the mechanical parts and can be detected by the heat energy generated in bearings etc. Furthermore if you subsequently bring the car to a halt you reduce its kinetic energy to zero since, by definition, the energy of motion is zero when there is no motion. When you stop the car its kinetic energy is converted to heat energy in the brake linings and tyres of the car.

This sequence of energy conversions shows that although the energy in the petrol performed the useful task of moving the car from one place to another, ultimately all the energy ended up as heat energy. Now if you could measure all the heat outputs, the heat in the exhaust gases, the heat generated in bearings, brake linings and so on, then, according to the law of conservation of energy, you would find that the heat energy output was exactly equal to the chemical energy input from the petrol. Although I don't know of anyone actually trying this set of measurements many simpler situations have been carefully monitored and in all cases the equality of energy inputs and outputs has been confirmed. In fact scientists and technologists have come to believe in this law so strongly that when they see energy apparently disappearing off somewhere they invent a new form of energy to account for the observation.

An important conclusion that can be drawn from this description is that the motor car has not *consumed* any energy. In fact as far as we know nothing can consume energy in the sense of making it disappear. What the car has done is to turn a very useful type of energy, namely the chemical energy in petrol, into a useless source of energy, some slightly warmed air. This is characteristic of all machines. Machines do not consume energy, they only degrade it from a useful form to a useless form. Of course in the process of degrading the energy the machine will perform some useful work, but ultimately the work will also be converted to heat, usually by friction.

In *energy* terms all machines are always 100 per cent efficient since the total energy input must, by the conservation of energy, equal the total energy output. So what does it mean when I say that power stations produce electricity with an efficiency of 25 per cent or that internal combustion engines are only 20 per cent efficient? The point is that technology is purposeful; machines and engines are built with some purpose in mind. If I construct a machine to convert coal energy into electricity, its efficiency is the electrical energy output divided by the coal energy input. If the desired output (electricity) is less than the input (coal) then the machine is said to be less than 100 per cent efficient. The energy that does not appear as a desirable output is counted as a loss, but it hasn't disappeared. In the case of an electric power station the 'loss' is the energy ejected up the cooling tower, in the case of motor cars the 'loss' is the energy thrown out in the exhaust gases. Thus a statement such as 'internal combustion engines are 20 per cent efficient' contains the implicit assumption that the energy in the exhaust gases is not useful and represents a loss.

But why should hot exhaust gases be useless? Why should the hot water or steam ejected from an electricity turbine be useless? Why can't we get the energy in these outputs to do useful work? To answer these questions we have to explore some of the consequences of the second law of thermodynamics, which is concerned with the transfer of heat.

Heat is a very special form of energy because unless you go to very elaborate lengths it will spontaneously flow away into the surroundings. If you put heat energy into an object its temperature rises, but when you stop putting heat energy into it, its temperature will quickly revert to that of the surroundings unless you carefully insulate it. This is because heat energy spontaneously flows from hot objects to cooler objects, but not in the opposite direction. A hot frying pan immersed in a bowl of water rapidly cools down and

## Where It Goes To

the water gets slightly warmer. Heat energy flows from the frying pan to the water. The heat flow stops when the pan and water are at the same temperature, when they are equally warm. This is all commonsense; the problems arise when we try to use heat energy to perform work.

Clearly, to use a source of heat as an energy source we have to extract the heat from the source. This means that we have to allow the heat to flow from the source to some cooler body. Heat will not flow if the temperatures are the same, so we have to establish a temperature difference. But once we have introduced a cooler body we know that a proportion of the heat energy will simply flow into it and not perform any useful work. If we prevent any heat from flowing into the cooler body then it might as well not be there and we have lost the temperature difference which is necessary to extract heat from the source. What all this means is that when we use a heat source as a source of energy we cannot convert all the heat energy taken from the source into useful work – some of the heat energy will simply end up in the cooler body. Of course if we could take the heat out of the cooler body and put it back into the heat source then this wouldn't matter – but this is exactly what the second law of thermodynamics says we cannot do. It doesn't say it's impossible to take heat from a cooler body to a hot one, it says that you're going to have to put work in to bring about the change. In fact under ideal conditions you would have to put exactly the amount of work into putting the heat into the hot source as you could recover by letting the same amount of heat flow from the hot source to the cooler body.

Since modern technology is almost totally dependent on the use of heat energy obtained from the combustion of fuels it would be nice to be able to estimate the efficiency with which heat can be converted to work in heat engines. To do this we have to quantify the ideas set out above. To start with let's just concentrate on one heat engine which will take an amount of heat energy $Q_1$ from a heat source and produce an amount of work $W$. The discussion above suggested that this engine must also reject some heat into a cold body. Let's denote this by $Q_2$, as shown in Figure 9. Let me also assume that this engine is as perfect as possible, in other words it has negligible friction. Then by the conservation of energy:

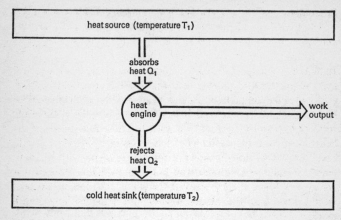

Figure 9 An idealized 'heat engine' operating between a source of heat at temperature $T_1$ and a cold sink at temperature $T_2$.

$$\text{total energy in} = \text{total energy out}$$
$$Q_1 = W + Q_2$$

This also means that the total work output ($W$) is equal to the difference in the heat input ($Q_1$) and the heat output ($Q_2$). Thus

$$W = Q_1 - Q_2$$

The heat engine has been designed to convert heat energy to useful work, so its efficiency is the ratio of the work output ($W$) and the heat energy input ($Q_1$). Denoting the efficiency by $E$ gives

$$E = \frac{W}{Q_1}$$

It is now possible to substitute the previous expression for $W$ into this equation to give

$$E = \frac{W}{Q_1} = \frac{Q_1 - Q_2}{Q_1} = \frac{Q_1}{Q_1} - \frac{Q_2}{Q_1}$$

So finally

$$E = 1 - \frac{Q_2}{Q_1}$$

In this form you can see that the efficiency can only be equal to 1·0 (or 100 per cent) if $Q_2$, the heat rejected into the cold body, is equal to zero. To proceed further in estimating the efficiency of heat engines we have to have a way of estimating the value of the ratio $Q_2/Q_1$. In other words we want to know how much of the heat taken from the source ($Q_1$) has to be rejected into the cold body ($Q_2$). This ratio must have something to do with the temperature difference between the hot and cold heat sources (or reservoirs), since if they were at the same temperature no heat would flow and the engine would not produce any useful work.

This conclusion, that the efficiency of heat engines depends upon their operating temperatures, is of fundamental importance to an understanding of fuel use. Using a little more mathematics[1] it is possible to show that the ratio $Q_2/Q_1$ is equal to the ratio of $T_2/T_1$ where $T_2$ and $T_1$ are the temperatures of the heat reservoirs expressed in degrees Kelvin.[2] Thus the final expression for the efficiency of a heat engine is

$$E = 1 - \left(\frac{T_2}{T_1}\right)$$

To put this result to use we have to establish reasonable estimates of the temperatures $T_2$ and $T_1$. The upper operating temperature of

1. This derivation is explained in any standard textbook on thermodynamics. See for example *Heat and Thermodynamics* by M. W. Zemansky (McGraw-Hill, New York, 1957 and later editions). A useful discussion of energy conversion and thermodynamics is in *Energy Power and Society* (T100, CU20/21), The Open University Press, The Open University, Walton Hall, Bletchley, Bucks.
2. This is the absolute temperature scale. Degrees Kelvin are the same as degrees Centigrade, but the Kelvin scale has its zero at the absolute zero of temperature. Thus 0 K equals −273·2°C and 0°C equals 273·2K. To convert a temperature in degrees Centigrade to degrees Kelvin all you have to do is add 273·2 to the Centigrade number. Hence 10°C equals 283·2K and 100°C equals 373·2K.

a machine is limited by the mechanical properties of materials at high temperatures. Steel structures or mechanisms cannot continue to be useful much above a temperature of 600°C, equal to 873K. The lowest temperature for the cold reservoir is the temperature of the local atmosphere or river.[3] Thus for practical purposes the lower temperature will be about 5°C, equal to 278K. Hence for a heat engine that can be constructed from real materials and can operate in our environment the maximum efficiency is

$$E = 1 - \left(\frac{T_2}{T_1}\right)$$

$$E = 1 - \left(\frac{278}{873}\right)$$

$$= 0.68 \text{ or } 68 \text{ per cent}$$

Remember this is still an 'ideal' limit in the sense that we have not considered any friction losses or other energy outputs. Nevertheless the calculation shows that at least a third of the heat input cannot be converted to work. In real machines, such as electric turbines and car engines, only half this theoretical efficiency is achieved.

Having established some of the implications of the laws of thermodynamics we can now return to the main theme of this book, which is why we need fuels and how we use them. It is important to remember that the conservation of energy applies to all types of energy conversions, whereas the second law and the efficiency limit apply only to the conversion of heat to work. Other types of energy conversion can be accomplished with a theoretical efficiency of 100 per cent, although in practice this is never realized. For example a hydro-electric power station converts the gravitational energy stored in water behind a dam to electricity with an efficiency of about 88 per cent. Electricity itself can be converted to work in power machines with efficiencies of over 90 per cent. But

---

3. It is no use considering using a refrigerator to create a lower temperature since the refrigerator only produces a lower temperature by consuming fuel.

the chemical energy in petrol or coal can only be converted to work with an efficiency of 20–30 per cent. It is this difference in the efficiency of conversion to work which accounts for the 'usefulness' of different fuels.

This can be illustrated by comparing the overall fuel efficiencies of petrol and electric cars. What we want to know is the ratio of primary fuel input to useful work done by the car engine. We have already established (Table 6; p. 44) that the efficiency of the oil industry is 88 per cent and that of the electricity industry 25 per cent. The measured efficiency of conventional petrol-driven cars is about 20 per cent, since the engine is 25 per cent efficient and the gearbox and transmission systems 80 per cent. Thus for every unit of oil which starts in the ground 0·88 reaches the car and 0·88 × 0·20 = 0·18 (or 18 per cent) is converted to useful work in the car. If we assume that the electric car has a similar transmission system with an efficiency of 80 per cent and that the charging of batteries and electric motor are also 80 per cent efficient, then the electric car produces 0·64 units of work for every unit of electricity. However, for every one unit of primary fuel (say oil) we only get 0·25 units of electricity, so the overall efficiency of the electric car is 0·25 × 0·64 = 0·16 or 16 per cent. The best oil-fired power stations achieve an overall efficiency of 33 per cent and if this were the source of electricity then the electric car efficiency would be 0·33 × 0·64 = 0·21 or 21 per cent.

Thus, as is shown in Figure 10, the overall efficiencies of both cars is very similar. In energy terms the significant difference is that in the electric cars the conversion losses are centralized whereas for petrol cars the conversion losses occur in each separate vehicle. It is worth noting that if you owned an electric car you would have to purchase only 0·25 units of electrical energy to achieve the same result as someone with a petrol car purchasing 0·88 units of petrol energy.

In terms of work this shows that 0·25 units of electricity are as valuable as 0·88 units of petrol – a ratio of $3\frac{1}{2}$ to 1. This is why an industrial nation like the United Kingdom converts a significant fraction of its primary fuels to electricity. It is also part of the reason why we are prepared to pay twice as much for a kWh of electrical energy as for a kWh of petrol energy. However, in terms

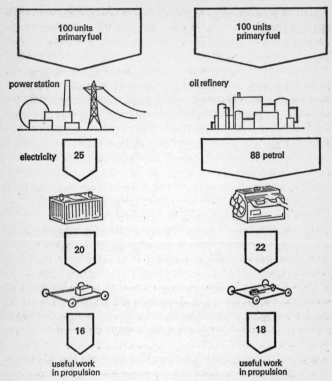

Figure 10 The flows of energy for petrol-driven and electric cars.

of other fuel uses the comparison gives different conclusions. About a third of all fuel used in the U.K. is used to provide low-temperature space heating, keeping homes, offices and factories warm. Electricity can be converted to heat with virtually 100 per cent efficiency, but coal and oil boilers produce heat energy with an efficiency of 75–80 per cent. Thus one unit of primary fuel will produce 0·25 units of heat in an electrically heated house but $0·88 \times 0·80 = 0·70$ units of heat in an oil-heated house. Thus for purposes of heating oil has almost three times the primary fuel efficiency of electricity.

This emphasizes that the 'usefulness' of different fuels depends upon the uses for which they are needed. For performing work electricity is about three times more efficient than oil, but for heating the situation is reversed. There are four broad categories of fuel use, namely low-temperature heating, high-temperature heating, performing work and special tasks associated with communication, lighting and control. Virtually any fuel, or indeed any source of energy, can be converted to low-temperature heat with a high efficiency. The reject heat from power stations or even weak sunshine can be used for this purpose. To obtain high-temperature heat, of the type needed to melt steel, you need to use either a high-grade chemical fuel (such as gas, oil or coal) or electricity. But there is no difference in the usefulness of oil and electricity for the provision of high-temperature heat – both are converted with high efficiency in the furnace. For performing work we have already compared oil and electricity and found that *overall* there is no difference in the conversion efficiency. For the specialized tasks such as operating telephones, television receivers and light bulbs, electricity is essential. It is possible to estimate the percentage of fuel used in each of these four categories by each of the sectors described in Chapter 4. For example in the domestic sector 65–75 per cent of the fuel use is for low-temperature space and water heating, about 10–15 per cent for cooking and boiling water (high-temperature heat) and about 5–10 per cent each for work and special applications. Breaking down each sector's use of fuels in this way, the fuel used in the U.K. can be broken down as in Table 11.

Table 11  *Fuel use by type of application*

|  | per cent |
|---|---|
| Low-temperature heat | 40–50 |
| High-temperature heat | 15–20 |
| Work (a) transport | 15 |
| (b) other | 20–25 |
| Special uses | 5–10 |

We can now return to the questions raised at the beginning of this chapter and begin to see the implications of the answers. The first question was 'why do we bother to waste primary fuel energy by generating electricity?' In fact if electricity is used for 'special uses', such as communications, the loss of fuel energy is unavoidable – there is no way to run your TV set on oil or coal except by converting the oil or coal to electricity. If electricity is used to perform work, such as propelling vehicles or bending metal sheets, there is no significant difference between the use of electricity and the direct use of oil or coal. Even though there is no significant difference in overall energy efficiency, electricity may often be used for work applications because it is so much easier to control. In contrast to these two types of fuel use, the use of electricity for providing either high-temperature or low-temperature heat represents a scandalous waste of fuel resources. The rejection of low-temperature heat up a cooling tower is particularly stupid when it is realized that a significant proportion of the electricity is then used to provide low-temperature heat in houses and offices. These factors will turn out to be significant when we examine the future of fuel supply and fuel conservation.

One of the major reasons for our carefully examining the conversion of fuels and the ways that they are used was to make it perfectly clear that all the fuels used eventually end up as low-temperature heat rejected into the atmosphere. The scientific law of energy conservation tells us that energy cannot be created or destroyed, so the total energy input in the form of fuels must equal the total energy output in the form of heat. A small fraction of our fuel use will be stored for a long time in the chemical energy of some of the products we make, but this is estimated to be much less than 1 per cent of total fuel use.

Although this conclusion is obvious if you accept the law of conservation of energy it may be worth following through some of the common fuel uses to show how this happens in practice. We have already examined the energy conversions associated with driving a car and, although there is a temporary conversion to energy of motion, we found that ultimately all the fuel energy ended up as heat. If you consider the fuel used in keeping your house warm it is obvious that, although there may be a temporary rise in internal temperature, once the heating is switched off the house cools down by giving its heat energy to the surrounding air.

The fuels used to operate a street lamp help keep the air above roads a little warmer. The lamps convert the electricity to heat and light (lamps get hot when operating). The heat is lost to the air immediately. The light temporarily aids vision on the road, but it is eventually absorbed by some surface which is slightly warmed by the radiant energy. Quite a lot of the energy which is used in making fertilizers is wrapped up in the chemical bonds of the fertilizer; the rest is lost as waste heat at the manufacturing plant. The energy in the chemical bonds of the fertilizer is partially released as heat when the fertilizer is absorbed by a plant and most of the rest released when the plant is eaten or allowed to decompose. Similarly some of the energy which is used in making a motor car is temporarily stored in the chemical composition and shape of the car, but most is lost as heat during the production process. The small quantity stored in the car is lost when the iron rusts or is crushed into scrap ready for the steel furnace again. If you follow any use of fuel through all the stages of production and use of end product you find that ultimately all the energy ends up as heat.

# 6 The Problem of Too Much

In this chapter I want to explore some of the environmental effects associated with the use of fuels. Briefly I want to explain why it is that the heat produced by fuels imposes a physical limit on the rate at which fuels can be used in the future. The existence of any physical limit to man's activities on earth has many serious implications. When the limit is on the rate of fuel use, the implications are extremely serious, so it is worth spending some time establishing the basis of the limit. The first part of the chapter is concerned with this, initially in terms of the simple relationship between heat input and temperature change, finally in terms of the very complex interactions which combine to give us our climate. This discussion will show that, while it is impossible to *predict* the exact consequences of putting a lot more heat into the atmosphere, we can identify factors which suggest that our climate results from a fairly delicate balance between very large forces. This makes it difficult to deduce the exact position of a climatic limit, but it does not affect the arguments which lead to the conclusion that a limit does exist somewhere. The last part of the chapter attempts to put this heat limit into the context of the United Kingdom and hence estimate the upper limit on fuel consumption here. This is obviously an important result, so it is worth taking some care to arrive at the best answer given the present state of our knowledge.

The first fact necessary in understanding why there should be a limit on the rate of use of fuels was established at the end of the last chapter. It is that all the energy in the fuels used ends up as low-temperature heat put into the atmosphere. The second step in the argument is that when you put more heat in something it will get hotter. This is a phenomenon which often occurs in our everyday lives. For example if you put a pan of cold water on a hot stove there is a flow of heat from the stove into the water, so the water gets hotter, its temperature rises. Similarly if you put a poker into a coal fire the temperature of the poker will rise; but it will not continue to increase for ever. The poker can only get as hot as the fire. When the fire and poker are equally hot, when their temperatures are the same, there is no net flow of heat between them. Then the fire and poker are said to be in 'thermal equilibrium', since any heat flow into the poker is balanced by an equal flow of heat out of the poker.

This example can also illustrate another important idea, namely that objects can lose heat by radiation. If a very hot poker is taken out of a fire, it is no longer in 'equilibrium' with its environment, which is now cool air. The poker therefore puts heat into the air until its temperature is the same as that of the air. There are two flows of heat involved. The first arises by its being in contact with cool air. The second is of more interest to us. It is the heat energy given out as radiation by the poker. If it is hot enough the radiation will be visible and will make the poker 'glow' red. The glow is simply light generated within the poker. Even when the glow is no longer visible the poker still loses energy in this way, only now the radiation is known as 'infra-red radiation'.

Instead of thinking about a poker let us consider a barren planet in space. This planet will receive a very large heat input, in the form of light, from the sun. Since the planet is surrounded by empty space the only way it can lose heat is by radiating energy out into space. Just as in the poker example the planet will reach a state of 'thermal equilibrium' in which the energy it receives from the sun will be balanced by the energy it radiates into space. If an object can lose energy only by radiation, then there are well-established laws of physics which describe the relationship between the surface temperature of the radiating body and the rate at which energy is lost by radiation. It is found that the rate at which energy is radiated is proportional to the fourth power of the surface temperature.[1]

Now any small increase in the energy input to this planet must

1. The 'fourth power of the temperature' means the value of the temperature (on the Kelvin temperature scale) multiplied by itself four times.

be balanced by an equal increase in the energy output. However, the only way the planet can increase its energy output is to increase the energy radiated, which can only come about by an increase in the surface temperature. The radiation equation (also known as the Stefan-Boltzmann law) can be used to relate the change in the rate of energy input (which must equal the change in the rate of radiating energy) to the surface temperature. If we denote the energy input per year by $E$ and the change in annual energy input by $\Delta E$ (delta-$E$) then this is related to the surface temperature change, $\Delta T$, and the average surface temperature, $T$, by

$$\frac{\Delta E}{E} = 4\frac{\Delta T}{T}$$

In words, the percentage increase in annual energy input equals a quarter of the percentage increase in surface temperature. Thus a 1 per cent increase in annual energy input would increase the temperature by 0·25 per cent.

If we apply this fairly simple description of radiation energy balance to the earth as a whole, we find good overall agreement. The annual energy input to the earth from the sun is about $1.5 \times 10^{18}$ kWh and to radiate this much energy would require the earth to have a surface temperature of 280K or 7°C.[2] This is a fairly good average for what we actually observe on the earth's surface.

Now heat energy put into the atmosphere by the use of fuels constitutes an additional heat input to the earth in the sense that it is not normally part of the heat flows which comprise the earth's thermal equilibrium. Oil, coal and gas represent a store of solar energy that was incident on the earth millions of years ago. It took about a million years of accumulating fixed solar energy to produce 1,000 million tons of coal – the amount used in six months by modern man. Nuclear fuels are also a form of stored energy, but they were formed during the evolution of the earth from its primeval state. By using these stores of energy, man is putting an extra energy input into the earth's atmosphere. Thus if ever man's rate of use of nuclear and fossil fuels reaches 1 per cent of the solar input, that is $1.5 \times 10^{16}$ kWh/annum, the radiation balance pre-

2. To convert from Kelvin to °C subtract 273·2. See footnote, p. 63.

dicts that the surface temperature should increase by 0·25 per cent of 280K, equal to 0·7K or 0·7°C.[3]

Before examining the validity of considering the earth as a 'lump of rock in space' there is one issue which is worth dealing with, namely whether it would be possible to invent a machine which could get rid of the excess heat without increasing the surface temperature or having some effect on climate. The short answer to this is that it is *in principle* impossible to make such a machine, since, to succeed, it would have to violate the second law of thermodynamics.

It is possible to construct a machine which would put more energy out into space than the earth loses naturally, but only by making the radiating surface of the machine hotter than the earth's surface. (Remember we found a definite relationship between energy radiated and surface temperature.) However, to make the machine radiate heat at a higher temperature than the earth's surface involves taking heat from the atmosphere up to the higher temperature of the machine. This transport of heat from a low temperature (atmosphere) to a high temperature (machine) involves performing more work and releasing more heat into the atmosphere – hence defeating the purpose of the machine.

It has been suggested that we can avoid altering our climate by ejecting a significant proportion of our waste heat into the deep oceans. This idea is based on a design of coastal power station which pumps its cooling water down to the ocean depths. The attraction of this proposal is that it would take a long time to raise the temperature of the oceans significantly and in that time we could find some other solution to the problem. However, hot water has the nasty habit of rising when placed in a lot of cold water. The effect of pumping a lot of hot water into the deep oceans would be to create a significant up-current of warm water from the deep ocean to the surface. It has been proposed to use this up-current as a giant fish-farm, since it would carry nutrients from the deep ocean. Unfortunately the up-current would also carry a lot of the heat that was supposed to be stuck at the bottom of the ocean and once in the surface waters the heat would enter the

3. Since the divisions on the Kelvin and Celsius temperature scale are the same, a *change* in temperature of 0·7K equals a change of 0·7°C.

atmospheric system. Thus this is not a way of making the heat disappear. (The proponents of the scheme have subsequently recognized this and have concluded that the major obstacle to future progress is solving the problem of heat release.)

Other schemes are being proposed to try to solve this heat-release problem, and it is certainly attracting more and more attention. To evaluate any of these schemes for yourself all you have to do is apply the laws of thermodynamics described in Chapter 5. The first tells us that energy cannot disappear but only be converted from one form to another. A proposed scheme must therefore either try to store the heat energy somewhere or to eject it out of the atmosphere somehow. Either of these proposals involves 'transporting' the heat from one place to another and probably from one temperature to another. Moving heat geographically will involve work (in pumping etc.), which will release heat into the atmosphere, and of course heat losses. According to the second law, trying to take the heat to either a high-temperature machine for ejection, or to a high-temperature store, such as deep rock foundations, will involve dissipating as much heat into the atmosphere as the heat stored or ejected. These are the reasons why I believe that this limit on man's activities cannot be avoided in principle.

Now we have to establish that this heating effect does apply to the earth and that it might occur within the foreseeable future. Clearly, if it is going to take several hundred years to reach a significant level of fuel consumption, we can leave the problem of getting rid of excess heat to future generations. So what we want is an estimate of the time-scale involved. For the moment let's see how long it would take for world fuel consumption to reach the 1 per cent solar level on the basis of present trends. It must be stressed that extrapolations such as these are *not* predictions, they are simply a device for estimating the time-scale associated with the problem. For instance *if* I could show that we would run into climatic problems in twenty years' time, and *if* everyone believed the analysis, then presumably fuel consumption would be reduced so as to avoid the problems. Thus my extrapolations are made with a view to being *self-defeating*; I neither expect nor want them to come true.

# The Problem of Too Much 75

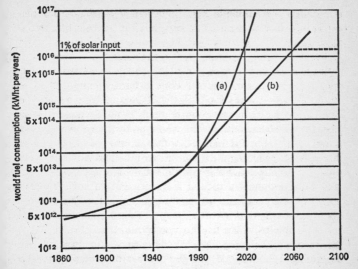

Figure 11 Historical trends in world fuel consumption. Note logarithmic vertical scale. (Sources: Putnam, 1952 and *U.N. Statistical Yearbook*.)

Figure 11 shows the trend in world fuel consumption since 1860. The vertical scale is a logarithmic scale, which means that equal divisions represent equal *multiples* of fuel consumption. Thus if fuel consumption were doubling every ten years (that is, growing exponentially with a doubling time of ten years), every ten years the fuel consumption would move the same distance up the vertical scale. The fact that the fuel-consumption trend is curving upwards slightly indicates that the rate of increase in fuel consumption is *faster* than exponential. Around 1900 the consumption of fuels doubled in thirty-eight years, by 1960 it was doubling every sixteen years and by 1970 it was doubling in about eleven years. If this decrease in doubling time continues, the fuel consumption will follow the curve marked (a) on the diagram. Alternatively the doubling time for fuel consumption could remain at eleven years, in which case the fuel consumption will follow the straight-line extrapolation, marked (b). These extrapolations show that, on the basis of present trends, global fuel consumption will reach the 1 per

## 76  Fuel's Paradise

cent solar level sometime between 2020 and 2060. This simple estimate of the time-scale associated with this problem puts it within our life times.

So far we have established that *if* the earth can be viewed as a 'lump of rock' then by about 2050 the global use of fuels could increase the earth's surface temperature by 0·7°C. I have also argued that there is no way of avoiding this. Now *if* the average temperature of the earth were to increase by 0·7°C, the polar ice-caps would melt (since ice and water co-exist only at one temperature, 0°C). *If* this happened, the level of the oceans would rise by about 100m (300 ft), which could flood vast areas of land. So the problem could be serious and within our time-scale. However, the earth is NOT a barren lump of rock. The thin skin of air and water surrounding the rock, the ocean–atmosphere system, significantly changes the situation. There are many interactions in this complex system, some of which tend to reduce the effect of heat inputs, others of which tend to amplify the effects. Climatologists are beginning to understand how the ocean–atmosphere works and the factors which influence its operation. However, our present level of understanding is not yet good enough to predict what would happen if the conditions changed. Although we cannot make predictions in this area it is useful to look at the operation of the atmospheric system so as to appreciate its complexity and the factors which affect the climate. One of the conclusions from this type of investigation is that while the atmosphere might reduce the significance of the heat released by fuel consumption there are other factors which may make the climate more sensitive to fuel consumption. But our first step is to understand the flows of energy in the atmosphere.

The incoming radiation from the sun is either reflected back into space, absorbed in the atmosphere or absorbed on land or sea. About 33 per cent of sunlight is reflected back into space, mostly from the tops of clouds. A further 22 per cent is absorbed in the atmosphere, leaving only 45 per cent reaching ground level. This is an average figure. On a clear day in mid-summer something like 80 per cent of the incident sunshine reaches ground level and on a cloudy winter day less than 20 per cent. Averaged over the year and over all weather conditions 45 per cent of the solar input reaches

The Problem of Too Much 77

the ground. This division of the energy input from the sun is represented on the flow diagram shown in Figure 12. The quantity of solar radiation reflected is determined by the reflection coefficient, shown as $a_1$ (alpha-one). The solar radiation absorbed by the atmosphere is determined by $a_2$, the absorption coefficient. The present average values of these coefficients are 0·33 for $a_1$ (giving 33 of the 100 units input being reflected) and 0·33 for $a_2$ (giving 22 units absorbed for 66 units input).

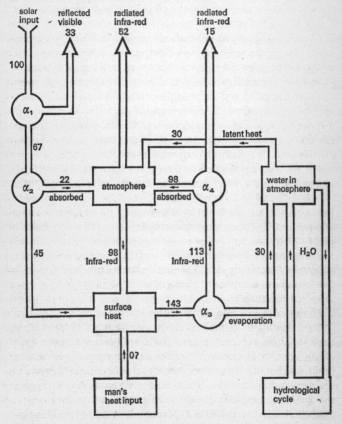

Figure 12  Flows of energy in the atmosphere (see text for explanation).

Also shown in this diagram are the flows of infra-red radiation between the earth and atmosphere and the flow of energy carried by water vapour. There are 98 units of infra-red radiation which are trapped in the circulation between the surface and the atmosphere. This flow of energy is trapped because only a fraction (determined by $a_4$) of the radiation from the surface passes through the atmosphere into space. Most of the infra-red radiation from the surface is absorbed by water vapour and carbon dioxide ($CO_2$) in the atmosphere and subsequently radiated back to the ground. This has the effect of making the surface temperature higher than it would be with simply 45 units of solar radiation. This is sometimes referred to as the 'greenhouse effect', since it is analogous to the way in which infra-red radiation is trapped by the glass of a greenhouse. Glass transmits sunlight, allowing the plants in a greenhouse to be heated, but the glass also absorbs most of the infra-red radiation given off by the warm plants. Thus infra-red radiation is trapped in the greenhouse and it is significantly warmer inside the greenhouse than outside. In effect water vapour and carbon dioxide provide a glass cover over the lower atmosphere.

The total energy flow on to the earth's surface is 143 units; 45 units of sunlight plus 98 units of infra-red from the atmosphere. This is balanced by 143 units of energy outflow, 113 units of infra-red radiation and 30 units by the evaporation of water into the atmosphere. The atmosphere receives a total of 150 units of energy input, 22 by absorption of sunlight, 98 by absorbing infra-red radiation from the surface and 30 from the condensation of water vapour into water droplets. The total output of the system, the output into space, comprises 33 units of reflected sunlight, 52 units of infra-red radiation from the atmosphere and 15 units of radiation from the surface, giving a total output of 100 units.

This simplified description of the energy flows in atmosphere shows that, as expected, there is an overall balance of energy. If man's input to the system were to become comparable to the solar input, we should have to incorporate this extra input in the analysis and we would find that it would lead to an equivalent increase in the total radiation into space. However, the main point I want to establish is that this would not come about by simply changing one or two of the numbers in the flow diagram – it may lead to changes in *all* the numbers. This is because there are a lot of interactions between components of the earth–atmosphere system which mutually affect one another and affect the partition coefficients (the $\alpha$'s in the diagram). A few examples of these interactions will show what I mean.

The ice cover of the polar regions of the earth are sensitive to small changes in the world climate. At the moment the ice-caps decrease the heat flow into the polar atmosphere by reflecting most of the incident sunlight and by insulating the air from the warmer water beneath the ice. In contrast ice-free water absorbs most of the sunlight incident on it and transfers a significant amount of heat to the atmosphere by conduction and convection. This difference between the properties of ice and water means that if the polar ice-caps *start* to melt they may melt away altogether under their own dynamic. The point is that if an area of ice is melted then the same area of ocean is created. But the ocean will now absorb the sun's radiation where previously the ice cover reflected it. Thus the heat input to the region will be increased. This increase in heat input will raise the temperature and cause a bit more ice to melt. In its turn the melting of more ice will further increase the heat input to the region and melt still more ice. This is an example of a positive feedback situation where a small initial change is amplified and brings about further changes.

Other interactions in the atmosphere exhibit negative feedback, which means that an initial change affects the system so as to partially cancel the initial change. For example an increase in surface temperature will increase the rate of evaporation from oceans, rivers and plants. This will increase the amount of water vapour in the atmosphere and hence increase the total cloud cover. However, clouds reflect a significant amount of sunlight incident on them and absorb most of the rest. Thus an increase in cloud cover will decrease the solar radiation and lower the heat input to the surface. This lowering of heat into the surface will partially cancel the rise in temperature which initiated this sequence of changes. The overall effect of increased cloud cover is very difficult to evaluate, since it will not only decrease the solar radiation reaching the ground but also increase the amount of infra-red radiation trapped in the 'greenhouse'. (In terms of Figure 12 an increase in cloud cover will

## 80  Fuel's Paradise

increase the value of the coefficients $a_1$, $a_2$ and $a_4$, but exactly how depends upon the type of cloud, its height etc.)

There are three factors which could substantially modify the pattern of heat flow in the atmosphere. The first is the heat release described at the end of Chapter 5. The second is the carbon dioxide generated when fossil fuels are burnt. The third is the dust ejected into the atmosphere when fuels are burnt and when the land surface is stripped of vegetation for certain periods of the year. The carbon dioxide ejected into the atmosphere enters into a complex series of cycles in much the same way as the ejected heat enters the system of heat flows of the atmosphere. A careful analysis of the

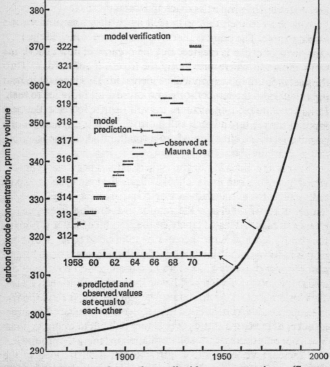

Figure 13 Changes in carbon dioxide concentration. (Source: Machta, 1971).

flows of carbon dioxide in the atmosphere–ocean system shows that it is removed from the atmosphere by a net transfer to the deep oceans. However, the rate of removal is significantly slower than the rate at which we put carbon dioxide into the atmosphere. Thus the concentration of carbon dioxide in the atmosphere is steadily increasing. The verification of an elaborate model of carbon dioxide cycles is shown in Figure 13.

Like carbon dioxide, dust does not stay in the atmosphere indefinitely. Dust particles either fall out of the air under their own weight or are washed out by rain. There is a very large natural concentration of dust in the atmosphere due to the occurrence of volcanic eruptions. Figure 14 shows the variation in annual atmospheric dust over the past 120 years. The large peaks are associated with volcanic erruptions; the rising solid curve is the estimate of man's annual contribution.

In one sense carbon dioxide and dust produce opposite effects in the atmosphere. An increase in carbon dioxide concentration will increase the 'greenhouse' effect in the lower atmosphere and it has been estimated that every 10 per cent increase in $CO_2$ concentra-

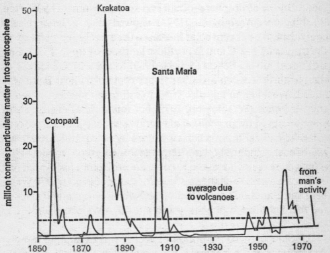

Figure 14  Time variations of dust put into the atmosphere (Mitchell, 1970).

tion will lead to a 0·2°C increase in temperature. An increase in dust in the upper atmosphere will increase the reflectivity of the atmosphere. An annual dust content of 45 million tons is estimated to cause a 1°C reduction in temperature as a result of the increase in reflected sunlight. (For a comprehensive review see Lamb, 1970.) However, dust in the atmosphere is also important in the formation of water droplets from water vapour. Water vapour can only condense to water around some suitable nucleus, usually a small dust particle. Thus an increase in dust in the atmosphere could increase the rate of formation of water droplets and hence increase the cloud cover. As indicated earlier the precise result of increased cloud cover is unknown, but it would certainly reduce the sunlight reaching the ground and probably change the patterns of rainfall.

As I indicated at the beginning of this discussion our understanding of the operation of the atmosphere is not sufficient for anyone to make firm predictions about any of the effects we have considered. However, all the effects do have the potential of significantly changing the global pattern of climate.

This is borne out by various attempts to use very large computer models of the atmosphere–ocean system. One estimate by Warren Washington (Washington, 1971) showed that a heat release equivalent to 5 per cent solar increased the average temperature in the tropics by 1–2°C and in northern latitudes by up to 8°C. Using a different model, Sellers (Sellers, 1969) found a 27°C increase at the North Pole when a similar (5 per cent solar) heat input was added. Washington has also tried to estimate the effect of much smaller inputs (Washington, 1972) but found that the changes in climate caused by an *addition* of energy input equal to 0·2 per cent solar were equal to the changes caused by an equivalent *subtraction* of energy input, showing that even this enormous model of the atmosphere is not adequate to solve the problem. This has led one eminent scientist to change his views of this problem. In 1970 Weinberg and Hammond published a paper which essentially dismissed the thermal limit argument, yet in late 1974 Weinberg wrote:

'Until recently consumption of energy was expanding rapidly. At the moment the increase is at a slower rate. However, there are great unsatisfied wants in many lands. When it becomes feasible to produce larger amounts of energy the former rate of increases might be resumed and even exceeded. Man now produces energy at the rate of $500 \times 10^{17}$ calories per year. This represents 1/20,000 of the total energy received by the earth from the sun, 1/5,000 of the total energy received by the earth's land mass. Man was increasing his production of energy by about 5 per cent a year; within 200 years, at this rate, he would be producing as much energy as he receives from the sun. Obviously, long before that time man would have to come to terms with global, climatological limits imposed on his production of energy. Although it is difficult to estimate how soon we shall have to adjust the world's energy policies to take this limit into account, it might well be as little as thirty to fifty years.

'Unfortunately, the science of climatology is unable to predict the ultimate consequences for the earth's climate of man's production of energy. At what rate of energy production would the ice caps melt? Will the carbon dioxide or dust thrown into the atmosphere by the burning of fossil fuel threaten the stability of the weather system? How does the geography of man's energy production affect weather in various parts of the world?

'Some attempts to answer these and similar questions have been made, for example, by computer modelling at the National Center for Atmospheric Research. Not enough is known to place too much confidence in such studies; yet answers to these questions may eventually dominate long-term energy policy. In the absence of such answers, how can we formulate intelligent policy?

'Two things should be done. First, climatologists should recognize the profound implications of this question and do the basic research in global modelling, in the dynamics of atmospheric circulation, and in increasing our general understanding of our global climate so that, say twenty years from now, we can base our energy policy on a much sounder understanding of this limit than we now possess.

'But this is not enough. The problem of global effects of energy production, like so many long-range environmental problems, is everyone's problem, and therefore no one's problem. I propose, therefore, that an institute (or even institutes) of climatology be set up with a long-term commitment to establishing the global effect of man's production of energy. Such an institute should be

## 84  Fuel's Paradise

assured long-term stability, since the question is a long-range one that simply will not go away. The institute would naturally serve to focus the efforts of smaller groups of climatologists, working on more general, basic aspects of climatology; but the institute itself would also contribute to our general understanding of the dynamics of the world's climate.

'I would hope that as part of our new appreciation of the necessity for truly long-range planning in energy, a strong, long-term effort along the lines I suggest will be launched.'

Personally I do not think that the global effects of fuel use will prove to be as constraining as local effects. This is simply because fuels are used in only a few areas of the world and the pattern of heat release is far from uniform. For example, the area of Greater London currently rejects heat into the atmosphere equivalent to 18 per cent of the solar input. The 4,000 square mile area of the Los Angeles basin currently dissipates heat equivalent to 5 per cent of the solar input and this is projected to rise to 18 per cent by the year 2,000 (Lees, 1970; S.M.I.C.; S.C.E.P.). These local heat outputs compare with the present global average of about 0·007 per cent solar. As we should expect from our earlier discussions these high heat outputs in local areas significantly alter the local climate. The centre of London is between 3 and 7°C warmer than its surroundings and there is a significant increase in cloud cover down-wind from the Los Angeles basin. In 1967 the temperature of San Francisco was 10°C higher than its immediate surroundings and there is evidence to suggest that more rain falls on cities than areas immediately up-wind or down-wind from them (Peterson, 1969). More recently the origin of a whirlwind in East Anglia has been traced back to the time when a cold front moved across the oil-refinery/power-station complex of the Thames estuary (Griffiths, 1974).

These examples illustrate that although our *global* fuel consumption is much less than 1 per cent we are beginning to observe local effects. These local climatic effects are an early warning system and will (I hope) stop us heading into much larger-scale climatic changes. This is important, since it means that we should try to relate the U.K.'s fuel consumption to both a global and a local climatic limit.

## The Problem of Too Much 85

The first step in discussing the heat limit for the U.K. is to compare our present fuel use with the annual solar input. To do this we have to calculate the annual input of solar energy, taking into account the average cloud cover, the change in length of day and the average declination of the sun throughout the year. At the edge of the earth's atmosphere the sun's radiation gives a power input of 1·4 kW per square metre. This corresponds to the 100 units entering the top left of Figure 12 (p. 77). This diagram also shows the average percentage of sunshine reaching ground level as 45 per cent, corresponding to 0·63 kW per square metre. This is the sunshine input to a square metre at the equator. In the U.K. the average angle between the sun and the ground is 50°, so the input per square metre of land area is reduced, as shown in Figure 15.

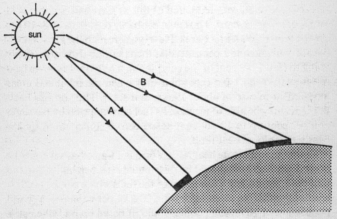

Figure 15  Both sunbeams (A and B) carry the same energy, but on the earth this energy is spread out over different land areas.

This latitude factor reduces the input to 0·43 kW per square metre for the U.K. However, this is the *peak* intensity, when the sun is highest in the sky. When the sun is on the horizon, at the beginning and end of each day, the input to a square metre of surface is zero. Averaged over the hours of daylight the solar input is half the peak intensity, that is, 0·21 kW per square metre. This value, which represents the daylight average, has to be divided by a further factor of two since over a year the length of daytime is 12 hours in 24 hours. So the equivalent continuous input to a square metre of land in the U.K. is half the daylight figure, that is, 0·11 kW/square metre. This is equivalent to 920 kWh per square metre per year (since there are 8,760 hours in a year). The total area of the U.K., including Scotland, Wales and Northern Ireland, is $244 \times 10^9$ square metres, so the annual solar input to the UK is $2,245 \times 10^{11}$ kWh/annum. In Chapter 4 we found that the total fuel input to the U.K. (counted on a heat-release basis) was $2,300 \times 10^9$ kWht/year; this is equivalent to 1·02 per cent of the solar input. Since the evidence suggests that the limit on fuel use should correspond to about 1 per cent of the solar input, the U.K. is at its heat limit.

There are, however, a number of other factors which have to be considered. In the first place only about 16 per cent of the world's surface is covered with land, and of this at least half is either arid desert or polar ice-caps. Thus man probably only lives on 8–10 per cent of the world's surface area. If everywhere man lives had a heat-release equivalent to 1 per cent solar then the global average is only going to be 0·1 per cent of the solar input. The global average heat release will reach 1 per cent solar only when the inhabited areas have a heat release of about 10 per cent solar. Thus on this basis the U.K. may be able to increase its fuel consumption by ten times (1·0 per cent up to 10 per cent solar) before contributing significantly to a global heat limit.

Rather than allocate heat release on a land-area basis it could be argued that the heat should be distributed on a per capita basis so that each person has a fair share of the fuels which can be used. A scenario which involved allocating 20 kW per capita to a world population of 10 billion people would, if based on an 'all-electric economy', produce a heat output equivalent to 0·6 per cent solar, which is close to the heat limit. At present the U.K. has an average fuel consumption of 5 kW per person, so if the limit is fixed at 20 kW/person the U.K. can increase its fuel consumption only by a factor of 4 (assuming that its population does not significantly change).

However, I have already argued that it is most likely that the limits on heat release will come about because of *local* climatic effects, not global effects. An important point about local climate changes is that they depend both on the quantity and on the distribution of heat release. Although the heat released over all the U.K. is equivalent to 1·0 per cent of the solar input the heat released in the 616 square mile area of London is equivalent to 17·8 per cent of the solar input to the same area. Furthermore the rate of growth in fuel consumption is fastest in the regions that have the highest density of fuel consumption. For example in the five years 1968 to 1973 the fuel consumption in London increased by about 21 per cent (4·2 per cent per annum), whereas for the U.K. as a whole the increase was only 8 per cent (1·59 per cent per annum). This disproportionate growth cannot continue for long, since after another sixty years at 4·2 per cent growth the fuel consumption of London would equal that of the U.K. as a whole. But the trend could continue for another twenty to forty years,[4] after which time the fuel consumption in London would be equivalent to 32–48 per cent of the annual solar input and the fuel consumption of the U.K. as a whole would be between 1·3 and 1·6 per cent of the solar input. A heat-release equivalent to 50 per cent of the solar input over an area of 616 square miles would certainly change the local climate substantially and probably produce noticeable effects for a hundred miles or so down-wind. This is the level of heat release at which I would draw the heat limit and where I expect local effects to be noticeable. (If the effects are noticeable at this level, the people around at that time will be demanding a reduction in fuel use.) This argument therefore suggests that the overall increase in U.K. fuel consumption can only increase by a factor of 1·5 on today's level.

Clearly the local climatic effects could be delayed if the physical distribution of heat release were changed. The factor of 1·5 deduced above corresponded to a limit of heat release equal to 50 per cent solar for London. If the disproportionate growth in fuel consumption in London were curtailed and other centres, such as the North-West or Midlands, allowed to come up to this type of level, the overall increase in U.K. fuel consumption could be twice the above estimate before any climatic effects were noticeable. This fairly generous estimate of a three-fold increase in total use is

4. This is because there is a positive feedback effect at work. As the temperature in London rises more people will install air conditioning.

below the limit deduced on the 20 kW per capita basis. Thus in future discussions I will assume that the limit on U.K. fuel consumption is between three and four times our present consumption, that is, in the range 7,500 to 10,000 × $10^9$ kWh/year.

# 7 Fuel Supply Efficiency

One of the questions which must have occurred to you is: 'Where on earth is all this energy going to come from?' At a time when the press is full of reports of the problems involved in trying to get enough energy it seems a bit crazy to be talking about a problem of having too much. Strangely enough I think that this paradox is a fair statement of the most serious problem which could arise in the foreseeable future. The present trends in fuel demand and the types of technical solutions that are being offered in response to shortages in supply seem to me to be heading us into a future where we may always be short of fuel and at the same time have too much energy. The link between these two problems is in the 'efficiency' of the fuel supply industries.

The important distinction which resolves the apparent paradox of too much energy and too little fuel is the distinction between primary energy and delivered energy. Earlier we looked at the breakdown of fuel use in the U.K. and found that for every 140 units of primary energy input to the U.K. only 100 units of useful fuel was actually delivered to consumers. Thus the total heat release associated with the 100 units of fuel used was 140 units. By examining the fuel technologies which are being considered for the immediate future I think I can show you that this ratio is going to get significantly worse within the next twenty to forty years.

The history of the fuel industries has been one of improving the efficiency of particular processes or technologies. For example, around 1900 the efficiency for generating electricity in the U.K. was about 8 per cent. Today the average efficiency is 25 per cent and the best efficiency close to 40 per cent. The trend is illustrated in Figure 16 and is continued into the future on the assumption

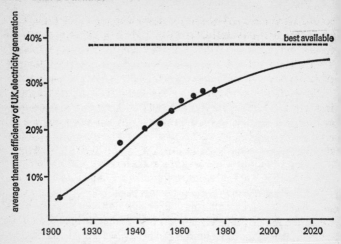

Figure 16  The average thermal efficiency of generating electricity in the U.K. From 1960 excluding nuclear stations. (Source: U.K. Energy Statistics).

that all that happens to the electricity industry is that old fossil-fired power stations are replaced with new fossil-fired stations. In a similar way the efficiency of the gas industry has improved. The recent use of natural gas in the U.K. has substantially improved its efficiency, since pumping gas from the North Sea requires considerably less energy than producing gas from coal.

However, these historical trends of improving individual efficiency may be offset by a decline in overall efficiency. The basic reason for the reversal is associated with the accessibility of resources. Initially fuel resources were very easy to exploit. The first coal mines were close to the surface and the first oil wells only needed a hole in the ground to go into production. As these easy sources were used up more difficult or more distant sources were used. However, by the time these more difficult sources were being used fuel technology had substantially improved, so the net efficiency of the fuel industries showed an improvement. Improvements in the design of oil refineries, delivery vehicles, electricity generating sets and so on more than offset the increased transport distances of bringing oil round the Cape or opening up deeper coal mines. Now these sources of primary energy are also being displaced by even more difficult sources. Electricity is to be generated from uranium, and oil is to be obtained from the bottom of the sea. The problem is that as the resources become more difficult to exploit our technology is approaching its limits of efficiency.

The simplest way to illustrate these trends and their implications is to consider the exploitation of material resources, for example copper ores. Around the turn of the century the average grade of copper ore was about 2·5 per cent copper. Now in South America the average ore grade is about 1 per cent and in the U.S.A., which

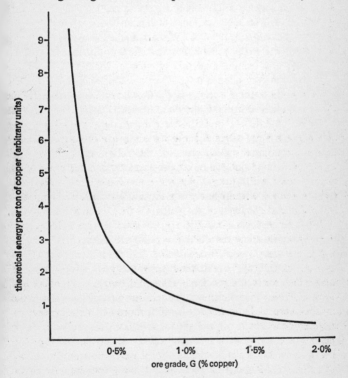

Figure 17 The theoretical energy needed to mine and mill enough copper ore to produce one ton of copper.

has exploited its resources more, the average is about 0·6 per cent. The significance of ore grade lies in the fact that a large slice of the fuel used to produce copper is consumed in digging ore out of the ground and transporting it to a mill, where it is crushed to a small size so as to allow separation of the mineral particles and 'dirt'. For a 1 per cent ore you have to mine, transport and crush 100 tons of dirt to produce 1 ton of copper, for a 0·6 per cent ore you have to process 167 tons of dirt, for a 0·1 per cent ore it's a thousand tons. Clearly as the grade of copper ore decreases the fuel used in the mining and milling operations per ton of copper will increase. In fact the increase will be proportional to the number of tons of ore which have to be processed, which is inversely proportional to ore grade (Chapman, 1974 (c)). We can pursue this analysis further so as to illustrate the role of technical efficiency. If the theoretical energy needed to mine and mill one ton of rock is $E_m$ then the theoretical energy needed to produce a ton of copper (excluding smelting and refining energy) is $E_m/G$, where $G$ is the ore grade. This type of relationship is shown in Figure 17. If we combine this information with the rate at which the grade of copper ore decreases with time, which depends on the rate of use of copper, we can obtain a curve which gives the theoretical minimum energy needed to obtain one ton of copper as a function of time. This is shown in Figure 18 (a).

Now since 1900 there have been considerable technical improvements in the efficiency with which all the machines – shovels, trucks and crushers – use fuel to perform their respective tasks. Thus over the same period the technical efficiency has been improving, as shown in Figure 18 (b). There is some evidence to suggest that this curve is flattening out, just as it is in the case of electricity generation efficiency (Figure 16), owing to theoretical limits. (In fact the efficiency of generating electricity is one of the important ingredients in the overall technical efficiency graph.)

The actual amount of fuel needed to produce a ton of copper at a particular time equals the theoretical energy divided by the technical efficiency. If this is worked out for the time variations shown in Figures 18 (a) and (b) then, as shown in Figure 18 (c), there is an initial decline, then a period of constant fuel/ton which is followed by a fairly sharp rise. This type of change in fuel input per unit out-

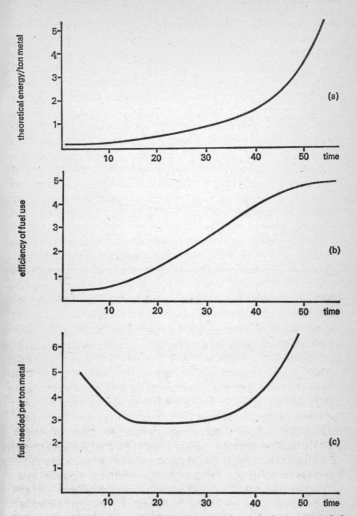

Figure 18 (a) As the ore grade declines the theoretical energy needed increases; (b) over the same time the efficiency of using fuels increases but saturates, so that (c), after an initial fall the fuel needed per ton of metal rises sharply. (Note arbitrary vertical scales).

## 94 Fuel's Paradise

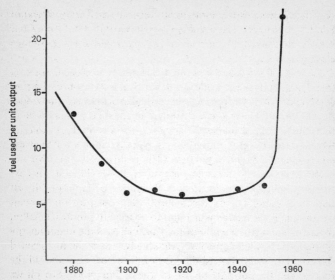

Figure 19  Evidence for the model shown in Fig. 18. (Source: Lovering, 1969).

put has already been observed for the U.S.A. (Lovering, 1969). Figure 19 shows the ratio of fuel consumed to material produced in U.S. mines since 1870.

This result has important implications if it is combined with the notion of a climatic limit on fuel use, since it implies that at some time in the future the climatic limit will force us to produce less and less materials as the grade of ore continues to decline. However, the point of this book is not to discuss materials policy but energy policy, and what we really want to know is whether this same type of relationship applies to energy resources. Here we run into a difficulty, because although it is pretty easy to define the 'grade' of a copper resource it is almost impossible to describe the 'grade' of an energy resource. As we have already seen in Chapter 5 different types of energy are not equally useful. More importantly, as technology changes so too does our definition of energy resources. Two hundred years ago oil was not a useful resource because no one knew how to refine it or use it effectively. Fifty years ago no one knew how to use uranium to produce energy. Today we do not know how to use the deuterium in the oceans, though we know that if we could make a fusion reactor the sea would be an enormous energy resource.

There is a good case for saying that there is no sense in looking for some directly measurable 'grade' of an energy resource. Instead we should be looking at something else. The oil that is found under the North Sea is just as 'concentrated' as the oil that was found in the Middle East or in Texas. However, there is no doubt that all these sources of oil require different resource inputs for their exploitation. There is no doubt that, as far as an automobile owner in Texas is concerned, Middle East oil is more 'difficult' or 'expensive' than Texan oil and that North Sea oil is more difficult than both these. The most significant difference, for our present discussion, is the energy which has to be used to produce a gallon of fuel *at the petrol pump*. For instance in Texas the major energy input needed to make the fuel available for use is the fuel needed to run the refinery. For Middle East oil we now have to add the fuel needed to transport the oil to Texas. For North Sea oil we have to add a fraction of the fuel cost of the oil rig and pipelines plus transport costs. So although in the oil industry itself the operation of refineries, ships, drills etc. has been made more efficient there has still been a net increase in the fuel cost of oil due to the increasing 'difficulty' of oil resources. (But in chemical terms the grade of oil has remained constant.)

This suggests that some parameter related to the fuel cost of fuels is the equivalent to ore grade in the case of copper. Note that since the 'fuel cost' represents the total primary fuel needed for the production of the fuel, plus its energy content, this is a measure of the total heat release associated with the provision of a unit of fuel. To date only approximate values of the fuel costs of fuel are available (Chapman, 1974 (a)). However, these do suggest a trend of increasing fuel costs. Table 12 gives the best estimates currently available for the future fuel costs of gas and petrol from a range of sources. Even though these numbers are likely to be revised by more thorough calculations now under way, it is unlikely that the revisions will change the direction of the trend.

There are two other trends in the fuel supply industries which

*Table 12 The approximate fuel costs of fuels*

| Gas | kWh/delivered therm gas |
|---|---|
| Town gas (prior to 1968) | 40 |
| Natural gas (North Sea) | 30 |
| Liquefied natural gas (Middle East) | 30–35 |
| 'Hygas' (from coal) | 40–45 |

| Oil | kWh/gallon petrol |
|---|---|
| Middle East | 53·6 |
| North Sea | 59·0 |
| Oil shale | 60–70 |
| Syncrude from coal | 60–80 |

will aggravate this ratio of heat release to delivered fuel. The first is the trend towards increased electrification. We will examine this in more detail in later chapters. For the moment it only needs to be emphasized that the trend towards nuclear power will increase the fraction of fuel delivered as electricity, which will substantially increase the heat release/delivered fuel ratio. The second trend is one which we will examine in some detail. It is the trend towards increased capital intensity in the fuel industries. This is a reflection of the increased 'difficulty' referred to earlier. However, it has an extra significance because a large capital investment also carries with it a substantial fuel investment. The trend towards more capital-intensive fuel sources means that the amount of fuel that has to be committed (burnt) to ensure future supplies is also increasing. This means that a larger and larger fraction of the 'delivered fuels' are actually burnt in order to provide future fuel supplies, that is, they do nothing towards satisfying the demand for fuels in copper mines, shops or homes. At the moment about 2·5 per cent of total U.K. fuel is used in this way (Chapman, 1974 (a)). However, if the current plan to develop the North Sea (at a total cost of £5,000 million) and the plans (prior to 1975) for increasing the number of nuclear power stations (at a total cost of £8,000 million) are assumed to take place in the next five years then the average fuel investment would have risen to about 4·5 per cent of total fuel use.

electricity will have to be used for heating. In the case of a nuclear power station we have found an increase in fuel supply (the electricity output) and simultaneously an increase in thermal and electrical fuel demands. It is reasonable to assume that the electricity generated is used to meet both these demands, since any other assumption requires some other consumers to change their fuel use. A more thorough discussion of this conventional problem occurs later in the chapter. For the moment it should be noticed that this convention, of using electricity to supply all the inputs, is certainly valid in an 'all-electric' situation but that this is a limiting case. On this basis the net output of a 1,000-MW station is, as shown in Table 14, equal to 523·4 MW. Assuming that the station operates for twenty-five years this means that the total net output of the station is $114{,}625 \times 10^6$ kWh of electricity.[1] This is a very handsome profit on the initial investment, which was $10{,}233 \times 10^6$ kWht.

*Table 14  Outputs of 1,000-MW SGHWR reactor*

| | |
|---|---|
| Installed capacity | 1,000 MW (v) |
| Electricity generated (continuous equivalent) | 620 MW (e) |
| Used by electricity boards, etc. | 23·25 |
| Distribution losses | 46·50 |
| Power needed to make up loss of heavy water | 3·43 |
| Power for refuelling (uranium from 0·3 per cent ores) | 23·40 |
| Total electrical output to final consumers | 523·4 MW |
| Over 25 years | $114{,}625 \times 10^6$ kWh (e) |

The first way in which we can put this data to use is to examine how the ratio of inputs to outputs changes as the grade of uranium ore changes. The most straightforward way to do this is to evaluate the net electrical output per tonne of uranium used in an

[1]. The calculation goes as follows. 523 MW equals $523 \times 10^3$ kW. There are 8,760 hours in a year, so $8{,}760 \times 25$ hours in twenty-five years. Hence total output equals $523 \times 10^3 \times 8{,}760 \times 25$ kWh $= 114{,}625 \times 10^6$ kWh.

SGHWR. The initial core of an SGHWR requires 643 tonnes of natural uranium and the reactor refuelling requires 126 tonnes/year for twenty-five years. This gives a total of 3,793 tonnes of uranium used by the reactor in its lifetime. The reactor produces an output of $114,625 \times 10^6$ kWh of electricity over the same period. However, to calculate the net yield of the reactor per tonne of uranium we have to subtract the fuel costs of constructing the station and of enriching the initial core. Since we are interested in the net yield of a nuclear system it is reasonable to assume that all the inputs to the station construction come from nuclear electricity, both the electrical and thermal requirements. (As before, this assumption would certainly be valid in an all-nuclear fuel economy.) Thus from this output of $114,625 \times 10^6$ kWh we have to subtract the constructional items in Table 12, using the values shown in the second column. We also have to subtract the initial core enrichment and fabrication but not the mining energy (since this will vary with the grade of uranium ore). This gives a net yield of $109,216 \times 10^6$ kWh of electricity, equivalent to $28 \cdot 8 \times 10^6$ kWh of electricity per tonne of natural uranium.

Now we need to compare this yield with the fuel costs of mining and milling uranium ores of different grades. The principles are exactly the same as for the copper-ore analysis described earlier in the chapter, only now we want to extend the analysis over a much wider range of ore grades. The easiest way to do this is to plot the fuel-cost/ore-grade relationship on a graph where both the vertical and horizontal scales are logarithmic. This means that if fuel cost is inversely proportional to ore grade then the graph is a straight line sloping from top left to bottom right. To extend the analysis over a wide range of ore grades it is also necessary to introduce another parameter called the 'stripping ratio'. This is the ratio of tons of overburden (which cover the ore body) to tons of ore. When the ore is fairly rich, stripping ratios are high (up to 50:1). However, very low-grade ores have stripping ratios of zero, since the ore itself is virtually dirt.

For a mine with a high stripping ratio the fuel cost of removing the overburden will be a significant fraction of the total fuel cost. If we denote the energy to move a ton of material by $E_d$ and the energy to crush a ton of ore by $E_c$, the energy cost of mining and milling (either copper or uranium) is represented by

$$1\ (SE_a+E_d+E_c)/G$$

where $S$ is the stripping ratio and $G$ the ore grade. Figure 20 shows the family of curves produced by this equation for copper mines. The earlier fuel-cost/ore-grade curve (Figure 17, p. 91) corresponds to a linear (not logarithmic) plot of part of the curve for $S = 4$. Figure 21 shows the same curves for uranium mining based on the analyses of several mines of 0·3 per cent ore grade (with a stripping ratio of 24) and one mine with an ore grade of 0·007 per cent and a stripping ratio of zero. It should be emphasized that the

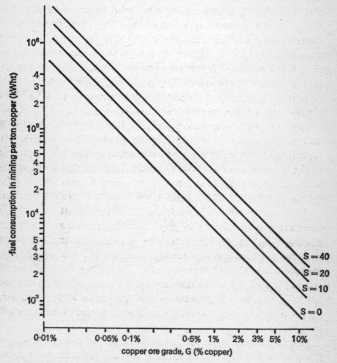

Figure 20 The fuel used in mining and milling copper ores as a function of ore grade (G) and stripping ratio (S). Note logarithmic scales.

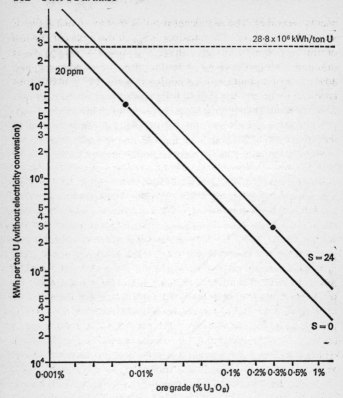

Figure 21  Fuel used in mining and milling ores (per ton uranium) as a function of ore grade and stripping ratio, S. Data from Chapman, 1974 (d)

calculations of mining and milling costs for the uranium graph do *not* include any 'fuel cost of electricity', that is to say the electrical input to these processes is counted directly without any conversion factor. The uranium curves are significantly higher than the equivalent copper curves because of the larger chemical and steam inputs needed in the processes used to separate uranium oxide particles from dirt.

I have spent some time establishing the basis of this calculation because it has important consequences for our later discussion of energy resources. The significant result is that at an ore grade of about 0·002 per cent (20 parts per million) the fuel cost per ton of uranium exceeds the net yield of $28·8 \times 10^6$ kWh(e) per ton of uranium. Thus as far as our present reactor designs are concerned any source of uranium with a grade lower than 0·002 per cent is not useful in the sense that it does not produce a net energy output. This rules out granite as a source of uranium, since the average concentration is 0·0003 per cent with a maximum concentration of 0·0016 per cent. In Chapter 10 we will find that eliminating very low-grade uranium ores has the effect of restricting the total number of nuclear burner reactors which can be built.

Having explored the uranium-grade relationship we can now also quantify the 'fuel-investment' problem discussed earlier. This is an interesting issue because although any one nuclear power station produces more fuel as output than it takes to build the station, the fuel input has to be made *now* in order to secure an output over the next twenty-five years. It is thus possible for a nuclear programme which is growing very rapidly to require more fuel as inputs to stations under construction than is available as output from the stations already built. This will only apply in periods of very rapid growth and the deficit will disappear as soon as the growth stops. Its significance is that since most industrialized countries are presently short of fuel and cannot afford activities which increase their present fuel consumption this investment problem imposes a limit on the maximum rate of growth.

To find this maximum rate of growth we shall start by looking at the time variation in fuel inputs and outputs of one station and then extend the analysis to two growth programmes. The basis of the calculations is summarized in Figure 22, which shows the fuel input spread over a five-year construction period, a one-year delay while the reactor is brought up to criticality and then a fuel output spread over twenty-five years. Thus if the fuel cost of the reactor is $10{,}233 \times 10^6$ kWht (corresponding to the first column of Table 13, p. 98) then the fuel input is $2{,}046 \times 10^6$ kWht per year (equivalent to 234 MW(th)) during the construction phase.

Clearly any nuclear programme will run into an initial fuel deficit during the first six years, since it takes at least that long for any station to start to produce an output. However, as far as the

**104** *Fuel's Paradise*

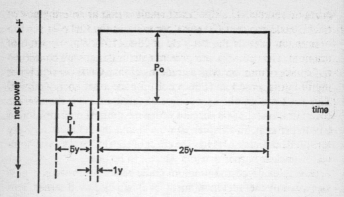

Figure 22 The power input and output of one nuclear power station.

U.K. is concerned, this initial deficit is only of academic interest, since the U.K. already has 5,000 MW of installed nuclear capacity. Thus the interesting question is not how much this has already cost but how fast can the U.K. increase its nuclear capacity from this point? I will consider two growth programmes, one with a doubling time of two years and the other with a doubling time of four years. To make a sensible comparison of these two growth programmes it is also necessary to impose a ceiling on the total nuclear capacity. For reasons explained in Chapter 10 I will impose a ceiling of 80,000 MW, sixteen times (or four doublings) larger than present capacity.

If the number of reactors is to double every two years, then with 5,000 MW in 1975 we need 10,000 MW in 1977, 20,000 MW in 1979, 40,000 MW in 1981 and 80,000 MW in 1983. But of course if this is the rate at which stations are *finished* they must be started at a similar rate six years earlier. Thus to have 40,000 MW available in 1981 we should have started at least this much capacity in 1975. Since there was 5,000 MW already finished in 1975 we should be building 35,000 MW of capacity – exactly seven times the capacity of stations finished. This simple calculation shows that for every one station finished we should have seven under construction in order to double the capacity every two years. This conclusion is derived more rigorously by mathematical analyses in Chapman

## Fuel Supply Efficiency 105

(1974 (d)) and Price (1974). The mathematical analysis shows that the 7:1 ratio applies throughout the period of growth.

Now one station produces an output of $4,585 \times 10^6$ kWh of electricity per year. To compare this with the inputs required by the seven stations under construction it is necessary to divide the inputs into electrical and thermal. Each station requires an annual input of $234 \times 10^6$ kWh(e) and $1,110 \times 10^6$ kWh(t), so the total demand of the seven stations is $1,638 \times 10^6$ kWh(e) and $7,770 \times 10^6$ kWh(t). If we allow the output of the one finished station to supply all the electrical input, this leaves $2,947 \times 10^6$ kWh(e) of output which can contribute towards the $7,770 \times 10^6$ kWh of heat also needed as an input. Since this need is for heat (principally in steel furnaces and cement kilns), the nuclear electricity is not more 'useful' than any other fuel and hence is subtracted directly from the heat required. This leaves a net thermal deficit of $4,823 \times 10^6$ kWht, which is the annual deficit for each completed 1,000-MW station. (Remember the 7:1 ratio applies throughout the growth period.)

There are two objections to this calculation. The first is that it is not possible to distribute the nuclear electricity to only those industries which are making nuclear power stations. In practice all electricity is purchased from a national grid, so most of the electrical inputs needed to build the stations will have come from fossil-fired stations. According to this convention the real input to the seven nuclear stations is $(4 \times 1,638) + 7,770 = 14,322 \times 10^6$ kWht, which gives a deficit of $9,737 \times 10^6$ kWht per 1,000-MW station finished. This convention assumes that the output of the nuclear station is evenly distributed to all consumers. This means that the *overall* efficiency of the electricity system is increased by the finished nuclear station, so that there is a slight reduction in the fuel cost of *all* the electricity used.

The second objection to the above analysis is precisely the opposite of this. It is based on a convention which says that the spare electrical output of the one finished nuclear station (equal to $2,947 \times 10^6$ kWh) is in fact 'worth' a lot more than the thermal inputs needed to build the seven stations ($7,770 \times 10^6$ kWht). This is based on the argument that if the $2,947 \times 10^6$ kWh of electricity were sold to consumers who wanted to perform *work* then it would be more 'useful' than $7,770 \times 10^6$ kWht of oil or coal. (This was explained in Chapter 5.) Thus the oil or coal previously used by consumers to perform work can now be used by the nuclear industry to build power stations and the nuclear electricity used to perform the work.

Both these objections are based on persuasive arguments, and all three conventions (the one used in the calculation and those of the two objections) are logically consistent. The trick is to try and guess which convention will be closest to the way people actually behave. Will all nuclear manufacturers purchase electricity from the grid? Will consumers who require work suddenly stop consuming oil and switch to electrical equipment? Both changes will happen *to a degree*. Some nuclear industries (particularly enrichment plants) may well get a special service from nuclear power stations – but not all of them. Some manufacturers who require work will stop using oil and convert to electricity – but not all of them. Thus on balance I think that the real deficit will lie somewhere between the $9,737 \times 10^6$ kWht/station (based on fossil-generated electricity) and the zero deficit (based on a change-over to electricity for work applications). Fortunately my own convention gives a value mid-way between these extremes, and is thus 'reasonable'.

We have now established that throughout a growth programme with a doubling time of two years the number of stations under construction is always seven times greater than the number of stations finished. Furthermore for each 1,000-MW station finished there is a net deficit of $4,823 \times 10^6$ kWht/year. Thus when there are ten stations finished the deficit is $48,230 \times 10^6$ kWht/year – and when eighty stations are finished the annual deficit is $385,840 \times 10^6$ kWht/year. The programme we are considering is shown in Figure 23, together with the changes in fuel supply and demand. The maximum deficit, which occurs at the end of the building programme, is equivalent to 17 per cent of present U.K. fuel consumption, which is not trivial. As soon as the growth stops the programme moves into profit and the annual output is equal to $366,800 \times 10^6$ kWh of electricity.

A similar analysis of a growth programme with a doubling time of four years produces the net fuel curve shown in Figure 24. Here the number of stations under construction at any time is only

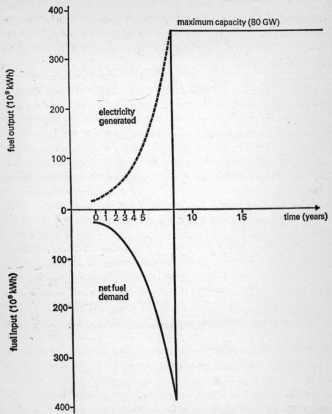

Figure 23  The fuel flows for a programme with a doubling time of two years.

twice the number of stations finished. This means that at any time (after the initial six-year deficit) the programme produces a profit of $1,897 \times 10^6$ kWh(e) per station finished. However, this programme takes a long time to reach the same final installed capacity. Under present conditions it seems preferable to the faster programme shown in Figure 23, simply because we cannot afford the deficit incurred in the faster programme.

Before leaving this analysis there is one further manipulation which we shall need when we want to incorporate these analyses into different energy policies. For reasons which will be made clear in Chapter 10, it turns out to be convenient to analyse energy policies in terms of 'primary fuel', not in terms of delivered fuel. Figure 24 gives us the delivered fuel output of a nuclear power programme with a doubling time of four years. In this programme each

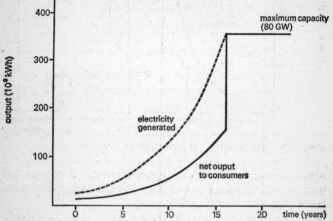

Figure 24  The outputs of a programme with a doubling time of four years.

1,000-MW station that is finished produces an output of $4,585 \times 10^6$ kWhe, of which $1,897 \times 10^6$ kWhe is available to consumers outside the nuclear industry. The primary fuel equivalent of one kWh of nuclear electricity is 3 kWh of heat,[2] since this is the amount of oil that would be needed to produce the same quantity of electricity. (It is also the heat released in the nuclear reactor.) Thus for a primary fuel input of $3 \times 4,585 = 13,755 \times 10^6$ kWht we have a 'useful' output of $3 \times 1,897 = 5,691 \times 10^6$ kWht of primary fuel. The difference between these numbers, $8,064 \times 10^6$ kWht, represents the extra heat released in nuclear reactors which does not contribute to consumer demands but which instead is converted to electricity and then used to build nuclear power stations. In dealing

2. It is interesting to note that, if we could find ways of using this heat directly, the preceeding analysis of nuclear stations would have to be based on a different convention.

## Fuel Supply Efficiency 109

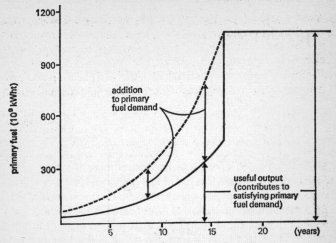

Figure 25  The primary fuel equivalent of figure 24.

with energy policies which use nuclear power, this primary fuel used in building has to be added to the primary fuel demand of consumers. This is shown in Figure 25.

Now one of the reasons why I think that this 'fuel investment' is a significant factor is that while the nuclear programme is growing fast the ratio of heat release (equal to total primary fuel used) to delivered fuels is a lot worse than at present. Although this contribution from the nuclear industry might cease quite quickly, there are other capital-intensive fuel industries which will also be growing in order to make up for the depletion of existing sources. For example, after the rapid build-up of nuclear burner reactors (the type analysed here) there could be a rapid build-up of nuclear breeder reactors. At about the same time we might have to start building expensive 'coalplex' plants to produce 'synthetic' oil and gas. Thus, although any *one* capital-intensive technology may contribute extra heat release only for a short time, we don't need one but a whole succession of capital-intensive fuel industries. This is why I think that we may run into the paradoxical situation of having too much energy (as waste heat) but not enough fuel (delivered to final consumers).

# 8 Energy Policy

The major aim of this book is to explore a number of options open to the United Kingdom in terms of fuel supply and fuel use. This is an exercise in energy policy, since we shall be looking at possibilities and constraints on the choices of future patterns and levels of fuel supply. Chapters 10, 11 and 12 will explore three types of future. The first is referred to as 'business-as-usual' and is based on the assumption that we just carry on doing what we have always done, but more so. The second type of future is referred to as the 'technical-fix' scenario and explores the ways in which technology can be used to reduce fuel consumption without substantially changing our style of life. The third future is a deliberately dramatic departure from our present life-style and is referred to as the 'low-growth' scenario. These three types of future, and their titles, are similar to those analysed by the Energy Policy Project (E.P.P., 1974) in the U.S.A. However, the American project was related to the particular constraints imposed by American resources, standards of living and so on. This book is an attempt to explore the options open to the United Kingdom, which have to satisfy the constraints of the U.K. and its standards of living. The purpose of this chapter is to try to spell out the constraints which are important for energy policies in the U.K.

Energy policy can be approached from two directions. The first direction involves examining potential fuel supplies and deciding what types of investments and decisions have to be taken in order to guarantee adequate future supplies. The second approach is based on the analysis of the demand for fuels. The future needs of industry and domestic consumers are estimated on some basis and the energy policy formulated on the basis of these demands. Clearly any realistic policy must combine both approaches. To be successful a policy must match total demand with total supply and the 'mix' of supply with the 'mix' of demand. It's no good having enough energy in the form of coal when people need fuels to drive motor cars. Furthermore any energy policy must be compatible with other aspects of life. It should not demand too much capital investment, nor should it decrease the level of employment. It should not be based on obsolete technology, nor on wild technological fantasies. It should take into account any trade implications and be flexible enough to allow for drastic changes in the world political climate. All these interactions between energy policy and other factors impose formidable constraints on possibilities.

The previous chapters have given us a framework which can be used to evaluate various aspects of energy policy. The concepts of fuel cost and the flow of fuel through the industrial system will enable us to evaluate future demands and the savings which could be made. The idea of the energy efficiency of the fuel industries gives us a direct link between primary fuel supplies and fuel demands. It also provides us with a link between energy policy and the climatic limit on the rate of fuel use. Since this is the ultimate long-term constraint on energy policy, perhaps it is the right place to start.

In Chapter 6 I argued that the climatic limit on heat release in the U.K. lay somewhere between three and four times our present fuel consumption, that is between 7,500 and $10,000 \times 10^9$ kWht/annum. Over the past twenty years the rate of growth of primary fuel consumption has increased by 1·87 per cent (as shown in Chapter 10). This is an exponential growth law with a doubling time of thirty-seven years. Thus if this trend were to continue we could reach the climatic limit in fifty to seventy years' time. However, in the last chapter we found two factors which could increase the rate of growth in primary fuel demand, although the growth in delivered fuels would be unaltered. These were the increasing 'difficulty' of fuel resources and the fuel investment problem. These could increase the rate of growth of primary fuel to 2·2–3·0 per cent per annum corresponding to doubling times of twenty-five to thirty years. This would take us to the climatic limit in forty to sixty years.

These time estimates may deceive you into thinking that as far as the U.K. is concerned we can leave the problem to future generations. Certainly forty to sixty years is a long way away, but by these times the U.K. fuel consumption will have to have stopped growing. To achieve this it is necessary to have started to slow the rate of growth much earlier. It is just about impossible for a modern industrial society *suddenly* to change any trend – all changes must be gradual. In fact the more slowly the change is made the less disruptive will be its effects. So even if the limit is eighty years away there are strong reasons for paying attention to it *now*.

It is important to understand something about the factors which

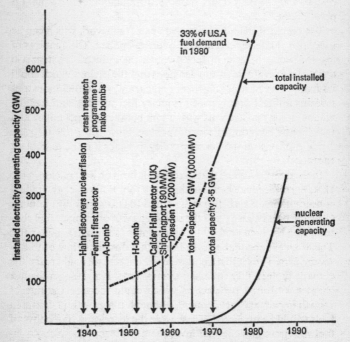

Figure 26 The historical and projected development of nuclear power in the U.S.A. (forecast from WASH-1139: 1974).

prevent a modern industrial economy from making fast changes, since these factors enter all types of policy formulation. As far as energy policy is concerned, probably the most important constraint is that imposed by technology. In simple terms it is impossible to bring about large-scale changes in technology in much less than twenty years. Although a technology may be 'invented' in a day, it will take three to eight years to get the technology scaled up, to get production lines equipped and prototype plants built and tested. With good luck it can then start to displace its predecessor by growing very quickly. Constraints on capital investment (similar to those discussed in connection with the energy investment of nuclear power in Chapter 7) mean that the maximum growth is 20 per cent per annum, corresponding to a doubling time of $3\frac{1}{2}$ years. Even at this astronomic rate of growth it will be at least fourteen to twenty years (four to six doubling periods) before the new technology has taken over a significant fraction of the total supply. This type of sequence is shown in Figure 26 for the growth of nuclear power in the U.S.A. Notice that there is about fifty years between the 'invention' and the time when technology is expected to make a significant impact on total supplies.

Part of this technological constraint on the rate of change is due to the fact that at any one time there will already be a viable system, based on an older technology, in existence. It does not make sense, in either economic or energy terms, to 'scrap' the existing technology overnight. Furthermore changing one part of an industrial system has many 'knock-on' and 'knock-back' effects.

A substantial shift away from oil towards nuclear electricity might require the introduction of electric cars. It would also require the construction of a larger-capacity transmission system as well as support technologies such as fuel enrichment and reprocessing. Even if we could invent a new type of housing which drastically reduced fuel consumption, it would take fifty to sixty years to replace the existing stock of houses. Furthermore, if the new-style houses required new materials, plants to produce them would first have to be designed and built. These 'system' effects of changes in technology provide another constraint on the maximum possible rate of change.

The last type of constraint on rates of change is both the most important and the most difficult to quantify. It is the constraint imposed by the norms and practices of our social system. It is not sensible to try to document all the social effects of different energy policies. Instead I will point to major types of effect and the constraints that these imply.

Ultimately all social effects reduce to changes in the activities or aspirations of individuals. For example certain types of policy may involve a technical change which will require certain individuals to change their employment, perhaps also their place of living. Changes in fuel taxes or the availability of fuels may change individual patterns of expenditure, with consequences reaching far outside the particular commodity taxed or in short supply. Sometimes the individuals concerned may not be within the U.K. system; for example changes in the attitudes and activities of certain Arabs can have a profound influence on U.K. energy policy.

All these matters give rise to three overall constraints on energy policy. The first concerns its relationship to the total productivity of our society. The total output of our industrial and social system is known as the Gross National Product (G.N.P.) and is found to be correlated with fuel consumption. The second overall effect is the effect fuel policy may have on unemployment. Unemployment can arise because of an overall decline in economic activity (which would also be shown by a decline in G.N.P.) or because changes in policy require new skills and a change in distribution of employment. The third overall effect is concerned with the trading position of the U.K. Energy policy can substantially alter the balance of payments of the U.K., since fuel imports presently account for over 25 per cent of all imports. Furthermore any technical changes introduced as a result of energy policies may change the relative prices of U.K. products compared to those of the rest of the world. Changes in prices can have large effects on our ability to trade.

When we come to look at each of the 'energy scenarios' in the next few chapters we shall have to examine each of these aspects of the energy policy in some detail. Before doing this it is worth outlining some of the general principles which relate energy policy to conventional indicators of economic activity. Perhaps the best-known relationship is that between the fuel consumption per

Energy Policy 115

capita and G.N.P. per capita for different countries. Figure 27 is a classic diagram illustrating this relationship. By comparing the

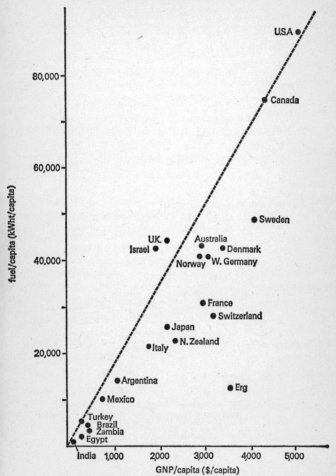

Figure 27  Per capita fuel use and gross national product for selected countries, 1970. (Source: U.N. *Statistical Yearbook*).

fuel consumption per capita and G.N.P. per capita for various countries the differences in population are removed. The popular interpretation of this diagram is that it shows a direct relationship between 'standard of living' (supposed to be represented by G.N.P./capita) and the fuel consumption per person. Some even go so far as to suggest that a reduction in fuel consumption *must* lead to a decrease in G.N.P. *and* a consequent reduction in standard of living. Although these are tempting interpretations they are gravely mistaken.

There are three kinds of mistake in these popular interpretations of Figure 27. The first is concerned with what is counted in a country's fuel consumption. The second is concerned with what is counted in a country's G.N.P. and how the G.N.P. is related to 'standard of living'. The third is concerned with the statistics of the situation and what can and cannot be implied by any diagram of the type shown in Figure 27.

Earlier we went to some lengths to establish the difference between 'energy' and 'fuel'. The distinction was made on the basis of 'usefulness'; a fuel is a technologically useful source of energy. One of the problems in compiling a diagram such as Figure 27 is that what is technologically useful to one country may be totally useless to another.[1] In practice the fuel consumption of different countries is evaluated on the basis of what is technically useful to the advanced industrial nations of the world. To illustrate why this is a questionable procedure we only have to notice that the energy content of all the food grown in the world is equal to half the energy contained in all the coal mined. The coal energy is included in the fuel consumption, but the food energy is not. In industrial countries, where the energy in the coal and oil used is about 100 times the energy in the food used, this is unimportant. In over two thirds of the world the principal source of energy (fuel?) is food, and inclusion of this in the fuel consumption/capita for India or China would roughly double the value recorded in the diagram.

The evaluation of a country's G.N.P. actually raises more prob-

1. Some economists have tried to incorporate 'efficiency' multipliers in their analyses and have been able to improve the correlation between G.N.P. and fuel. However since the multipliers are deduced from G.N.P./fuel data this is a self-justifying circular argument and of no physical significance.

lems than it solves. For an industrial nation, where actual production statistics are relatively easy to come by, the problems centre on how to count 'education', 'health services', 'police forces', etc. These activities do not actually produce a 'good' or 'commodity' which enters the 'market place'. In classical economics they do not contribute to 'gross national *product*'. However, they undoubtedly contribute to whatever is meant by 'quality of life' or 'standard of living', so if G.N.P. is supposed to represent these then some method of counting must be employed. In industrial nations this is done by counting a percentage of the 'gross turnover' of these service sectors, but there is no justification for this particular procedure.

In non-industrial nations the G.N.P. is evaluated according to the rules applying to industrial nations. Many authors have pointed to the absurdity of this and I doubt whether any economist would claim that a comparison of G.N.P./capita between industrial and non-industrial nations showed anything meaningful. In most non-industrial nations production statistics are not available and can only be estimated to an order of magnitude. Furthermore in these nations it is almost certain that the most important factors in 'quality of life' are *not* included in the conventional accounting of G.N.P. Produce grown for own consumption or services performed in exchange for subsistence are a major part of non-industrial life, but are excluded from G.N.P. accounts.

There are many smaller problems both in the technical evaluation of G.N.P. and in its interpretation. There is only one certainty and that is that G.N.P./capita provides an *extremely* crude measure of quality of life. When confronted with the detailed arguments most economists agree with this statement. However, they then point out that there is no other indicator of 'welfare' or 'quality of life' available and proceed to use G.N.P. as if it was a good measure. I urge you to interpret G.N.P. as a very bad measure of welfare.

The final fallacy involved in the popular interpretations of Figure 27 is concerned with the statistical nature of the diagram. The diagram shows the correlation between two variables (fuel/capita and G.N.P./capita) for a number of countries. It does *not* prove that there is any causal relationship between the two variables. It could be that there is a third variable, not shown on the diagram, that is itself causing the correlation. In fact a fairly convincing statistical case can be made that this third variable is 'motor cars per capita'. This also makes sense physically, since motor cars consume the type of fuel counted in the fuel/capita and the manufacture of motor vehicles is counted in the conventional definition of G.N.P./capita.

For me the most interesting aspect of Figure 27 is the wide *variation* between different countries. The trend of increasing G.N.P. and increasing fuel use is physically obvious (as discussed in Chapter 3). What is not so obvious is how countries like New Zealand, Switzerland and France can have slightly higher G.N.P.s than the U.K. with half the fuel consumption per capita. For my money this is far more important than the trend because it suggests that the U.K. might be able to simultaneously increase its G.N.P. *and* decrease its fuel consumption. This is in flat contradiction to the popular interpretation referred to earlier.

To begin to understand the reasons for the large variations in the

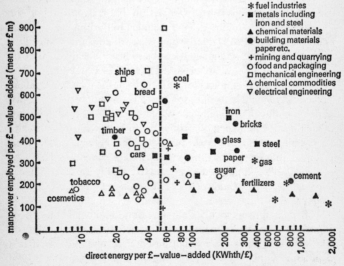

Figure 28   The use of manpower and fuel in some U.K. industries.

ratio of G.N.P./fuel use it is necessary to look on a finer scale than a total country. Figure 28 shows a plot of the manpower and fuel use in different industries in the U.K. Here the scales have been adjusted so as to give *manpower per £-value-added* and *fuel per £-value-added*. The reason for doing this is that the G.N.P. is the sum of all the '£ value added',[2] so the scales given are in proportion to the contribution to G.N.P. Thus an industry with a large value of fuel/£-value-added uses more fuel to make the same contribution to G.N.P. than an industry with a large ratio of manpower/£-value-added. Similarly an industry with a large ratio of manpower/£-value-added uses more men to make the same contribution to the G.N.P. than an industry with a small manpower/£-value-added ratio.

One immediate result of this disaggregation of fuel use is that a country of 'cigarette-makers' would require about two hundred times less fuel than a country of 'oil-refiners' to get the same G.N.P. In fact some of the differences between countries can be explained largely on the basis of different mixtures of the types of industries shown on this diagram. Generally primary materials industries require 100–1,000 kWh/£-value-added, manufacturing industries require 10–100 kWh/£.-v.-a. However, before this type of result can be put into any energy policy it is necessary to also take note of the manpower implication and then take note of the implications for trade. If we all decide to become car-insurance salesmen, we shall have to import the cars; if we want to stick at being motor-car manufacturers we may have to import the steel – and if we want to make steel and motor cars and do insurance, we have to import the fuels!

---

2. The 'value added' of an industry is the difference between its revenue from sales and its expenditures for materials and fuels. It is literally the value added to the material as a consequence of the processes of the industry.

# 9 A Prologue on the Future

To discuss such topics as oil supplies in 1995 or the growth of nuclear power in the year 2000, or even just to compile a graph which continues to the year 2010, is to open the door to all kinds of misinterpretations. Many people will interpret all or parts of the next three chapters as 'predictions of the future'. In fact rather than trying to predict the future I am trying to do exactly the opposite – to show the degree to which the future is under our control. My assumptions are that people are able, to a degree, to anticipate and solve problems. Thus these chapters should be viewed as preliminary explorations of the future and not in any sense as predictions.

In the process of exploring different types of future we shall be paying most attention to the problems and difficulties which are involved in trying to achieve each. This concentration on problems has led some people to view such exercises as being essentially pessimistic. This is again almost exactly the opposite of the truth. To illustrate why this is the case I want to point out a paradox which arose in the polarized debates between supporters of *Limits to Growth* and supporters of economists such as Professor Beckerman.

Essentially the debate between Professor Meadows (author of *Limits to Growth*) and Professor Beckerman (principal advocate of growth) hinged around what type of future lay in store for mankind. On the one hand Professor Meadows is reputed to have predicted a future fraught with resource shortages, food shortages, increases in pollution and horrendous population 'crashes'. On the other hand Professor Beckerman is reputed to have predicted a future full of abundance, including dramatic improvements

## A Prologue on the Future

in all aspects of life brought about by technological miracles. The paradox arises when these viewpoints are used as a basis for *action*.

Let's start by assuming that you accept Professor Meadows' description of the future. If you use this as a basis for action, presumably you will work towards either avoiding or anticipating the difficulties that he has pointed out. If enough people accept Professor Meadows' views, there is a very good chance that the difficulties and problems will be solved – in which case we end up living in precisely the type of future described by Professor Beckerman. On the other hand I could assume that you accept Professor Beckerman's view of the future. This does not require you to work towards solving or anticipating any problems. The difficulties are, according to the conventional economic view, supposed to work themselves out in the market place – with a little help from technology. If enough people accept this view of the future and simply carry on living their lives as before, essentially ignoring any potential problems, then you end up in precisely the type of disastrous future described by Professor Meadows. So who is the optimist, Professor Beckerman or Professor Meadows?

It seems to me that slogans like 'eco-doomster' and 'pessimist' or 'fixer' and 'optimist' are irrelevant to discussions about the future. If someone points out a potential problem they are being neither optimistic nor pessimistic, they are simply displaying an important aspect of human intelligence, namely the ability to anticipate events. If someone else dismisses the problem or – worse – ignores it, I don't think they are being optimistic or pessimistic, I think they are being stupid.

My own views about what will actually happen in the future change about as often as the English weather. However, one object of this book is not to persuade you to face the future either with a grin or with a frown, but to accept that there is no type of 'problem-free' future. All our options involve problems. Different types of future involve different types of problems, and whichever future you think we should be working towards, it is important to face up to the problems involved and try to find solutions.

There are two other features of the scenario explorations that follow which should be explained at this point, namely the time-span with which they are concerned and their numerical accuracy. The fairly detailed studies in each of the scenarios are restricted to the next twenty-five to thirty years and do not attempt to look further into the future. There are two reasons for this. First, as pointed out in Chapter 8, many social and technical factors constrain rates of change, so that it is possible to be fairly definitive about possibilities in thirty years' time. For example I am sure that no brand-new source of energy (such as fusion power) can make a significant impact on world or U.K. fuel supplies within thirty years. However I cannot make such a statement for the next fifty to sixty years. If fusion reactors were developed within the next twenty-five years, they could make an impact within a fifty-to-sixty year period. Thus by constraining myself to the next twenty-five to thirty years I am decreasing the degree of uncertainty involved in the study.

The second reason for looking only at a period of twenty-five to thirty years is that it corresponds to the time-scale over which 'long-term' policies can be expected to operate. The still longer term is then seen as being made up of a series of twenty-five-to-thirty-year periods. This is not unreasonable, since at any point in time the options open to you are largely determined by where you are at that time. If policies are implemented now, you will find yourself in a new position in thirty years' time. Your choices for the next twenty-five to thirty years are then constrained by the new position. It does make sense to cast a very long-term look at the direction in which your policies are taking you, but you cannot let events sixty years from now influence present decisions, since you don't know what is likely in sixty years' time.

The other aspect of the scenarios which should be borne in mind is that the numerical results obtained, and hence the conclusions drawn, are only as valid as my starting assumptions. Where these assumptions are obviously important they are discussed in the text. However, to avoid boring you with long lists, most of the smaller assumptions are simply buried in the numerical data. As far as I am aware none of these small assumptions significantly alters my conclusions. However, had someone else gone through this exercise they would probably have chosen different assumptions and come up with different numbers. This, and the fact that most of the statistical data on which the analysis is based is only accurate to 20 per cent, means that these scenario explorations should be viewed as 'first looks'. This point is taken up again in Chapter 13, when the relative merits and drawbacks of the three scenarios are discussed in some detail.

# 10 Futures I: Business-as-usual

If it is correct to suggest that, to a large degree, our future is under our control, then it is important to decide in which ways we want to control and direct our future. This means that before starting to explore any scenario we have to understand why anyone should want to try to achieve the particular type of future involved. This is important because the reasons for trying to achieve a particular kind of future will influence the way that problems are perceived and tackled. The reasons establish an order of priorities and they establish the prevailing political and social expectations of members of the future society.

Now obviously no matter how much anyone wants some kinds of futures they will not achieve them. You cannot make everyone a millionaire, nor send everyone to the moon, within the forseeable future. In Chapter 8 we saw some of the social and technical factors which constrain the maximum rate of change of methods of production etc. There are also other constraints, such as the attitudes of our trading partners, which have to be considered when deciding on a particular type of future. So our first task is to pull together the reasons, the constraints and the assumptions which together describe and justify a 'business-as-usual' type of future scenario.

The shorthand description of this scenario, 'business-as-usual', is a fairly succinct summary of the underlying reasons for advocating it. It means that the trends and aspirations which currently make our economic system function are assumed to continue. Firms are assumed to continue to try to grow so as to guarantee their own security and to increase their disposable income (retained profits). The major political and economic institutions are assumed to continue to exist and exert pressures in their traditional roles.

It is important to realize that this scenario is a deliberate choice and does not arise simply because people (institutions) do not know what else to do. The choice is based on the fact that our present social and economic institutions have arisen in response to past needs and, unless those needs can be catered for in some other way, the pressures to keep the institutions will be overwhelming. This is particularly true of economic growth. I do not want to get involved in the rather sterile 'growth v. no-growth' debate. However, it is essential to point out that a major social problem perceived by individuals and governments in the U.K. is the inequality of income, opportunity and influence. According to the conventional view this inequality cannot be alleviated by reducing the wealth of the better-off members of society. Conventionally the only politically acceptable way of redistributing income is to increase the total wealth and give marginally more to the less affluent. This is also expedient, since while everyone's real wealth continues to increase they can be persuaded to ignore the gross symptoms of inequality. (At least they can be persuaded that they have more to lose by rocking the boat and so had better keep rowing.) If the growth in real incomes were removed, everyone's expectation of a better tomorrow could be threatened and the basis of social co-operation would also be threatened.

Now over the past two years the U.K., like most other industrial nations, has experienced serious economic problems, not least in purchasing fuels. It might therefore seem that a business-as-usual philosophy is already out of date. This is not the case. The present difficulties are seen as a hiccup and most policies are still based on the idea of growth, either in company sales or in gross national product. Throughout 1974, government, industrial and trades-union spokesmen continually referred to 'overcoming the present difficulties' and returning to a growth economy. But the purpose of this chapter is not to try to decide how many people still adhere to this philosophy but to see whether, as a basis for future planning, it is a viable philosophy.

The most important assumption that underlies the business-as-usual philosophy is that the external conditions for continued economic growth continue to exist. Apart from the obvious political assumption that individuals do continue to want increased wealth and are not prepared to take it by force (political revolution) there are also assumptions about the rest of the world. By far the most important is that the rest of the world will continue to adopt a 'business-as-usual' outlook. This is especially true of the U.K.'s major trading partners. For the U.K. to be able to continue 'business-as-usual' it must be able to buy both food and raw materials in world markets and to sell products abroad so as to pay for the food and raw materials. A major change in the attitudes or economic conditions of one or two of our major trading partners could very seriously upset the U.K.'s trade balance and with it the prospects for growth. The evidence for this is not hard to find. The OPEC cartel resulted in a major change in world trade in 1973/4. The rise in the price of oil substantially altered the terms of trade of international markets and either caused or inflamed inflationary trends in all major industrialized nations. This resulted in a fairly serious economic recession in the U.K. in 1974,[1] with a consequent decline in a major industry (car manufacture) and a slight decline in G.N.P. in that year. The change in price of oil led to fairly serious difficulties in international affairs, with all the major industrial nations tripping over themselves to sell advanced technology to OPEC nations. It is possible that other underdeveloped nations with large stocks of natural resources will follow the OPEC example and use their indigenous resources to try to reduce the inequality between industrial and non-industrial nations. If they do, and if they are even half as successful as the OPEC nations, then the subsequent strains on balance of payments, international monetary funds and availability of credit will force most industrial nations to adopt a philosophy different from 'business-as-usual'. Thus this scenario *presumes* that our major trading partners will not seek to change the terms of world trade, and those advocating this type of future presumably consider the risk of being wrong on this matter to be less significant than the benefits to be gained by following a business-as-usual policy.

With this background we can now start to explore a 'business-

1. At the time of writing it is not yet clear whether this recession is over or whether the worst is still to come.

as-usual' (b-a-u) future from an energy point of view. One of the nice things about this scenario is that it is fairly easy to forecast the implications of policies, since they are, almost by definition, continuations of past and present policies. (I suspect that the ease of continuing a curve on a piece of graph paper has a lot to do with why most policy-makers adopt a b-a-u philosophy.) Our inquiries will start by examining the trends in fuel supply and consumption in the U.K. By adding some fairly simple assumptions about the U.K. population and the way people spend money we shall be able to deduce a demand for fuel based on a continued growth in G.N.P. Having got a fuel-demand curve we will then examine the various options available to supply the fuel. As you might expect this is the area in which this scenario runs into most difficulties.

There are three aspects of fuel supplies that interest us. The first is the total primary fuel consumption, the second is the mix of fuel sources used to supply this primary fuel and the third is the 'delivered fuel' obtained from the primary input. The historical trend in the U.K. primary fuel consumption and sources of supply is shown in Figure 29.

The actual primary fuel consumption is represented by the uppermost jagged line. The fluctuations from year to year are caused by factors such as the severity of the winter and the state of the economy. However, underlying these fluctuations there is a trend of steady growth since 1946. The trend is emphasized by the smooth dashed curve which is an exponential curve with a doubling time of thirty-seven years, corresponding to a growth of 1·87 per cent per annum. The mix of fuels which make up the total consumption is shown by the lower jagged lines, the contribution of any one fuel in a particular year being equal to the vertical distance between the lines above and below the fuel name. There have been a number of significant changes in the mix of fuels. Perhaps the most obvious is the decline of the coal industry, from a peak production of 210 million tons in 1955 to about 125 million tons today. More recently there has been a very fast growth in the use of natural gas as a primary fuel, from nothing in 1965 to 11 per cent in 1972.

In addition to these shifts in primary fuel supply there have also been substantial changes in the fuels used by final consumers. The importance of these changes can be gauged from Figure 30, which

128  *Fuel's Paradise*

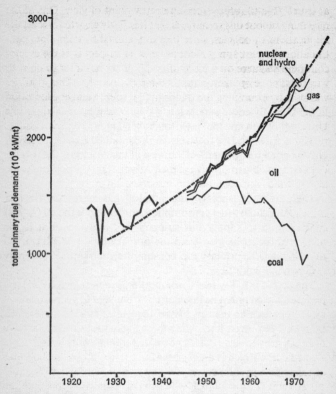

Figure 29  The historical changes in total primary fuel demand and the mix of fuels used. (Source: U.K. Energy Statistics).

shows the fuels delivered to domestic consumers (heat delivered basis) and the primary fuel needed to provide these domestic fuels. Although there has been virtually no change in the total heat content of the fuels delivered to our homes there is a very dramatic increase in our primary fuel demand. (In the terms introduced in Chapter 4, the lower graph shows the *fuel* delivered to domestic consumers, the upper graph shows the *fuel costs* of these fuels.) The large difference is due to a substantial shift in domestic con-

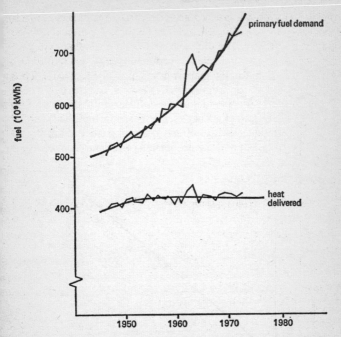

Figure 30 The heat delivered to domestic consumers and its fuel cost 1950-70. (Source: U.K. Energy Statistics).

sumption from coal to electricity.[2] Similar substitutions have taken place in other fuel-consuming sectors, with the result that a larger and larger fraction of the primary fuel input is used to generate electricity. The fraction of the primary fuel input used for electricity production is shown in Figure 31. This trend means that the overall efficiency of the fuel industries is decreasing. As pointed out in Chapter 5, provided the electricity is used for appropriate functions this does not represent an overall decrease in the efficiency of our

2. Historically this represented a substantial improvement in home comfort, since open coal fires are very inefficient. However modern coal heating systems, especially those using fluidized beds, are a lot more efficient. Using a modern system coal could provide as much 'useful' heat as can be obtained from an equivalent quantity of delivered electricity.

## 130  Fuel's Paradise

fuel use. However, as a larger and larger fraction of primary fuel is converted to electricity, electricity will be used for more and more inappropriate uses. The difference in the curves in figure 30 arise because of the growth in electric central heating (shown in Figure 32), which *is* a totally inappropriate use of electricity. Thus although the shift towards electricity *need* not lead to a decrease in efficiency of fuel use, in practice it *will*. It is therefore important to estimate the overall efficiency of the fuel industries. This can be done using the changes in electrical generating efficiency predicted in Figure 16 (p. 90) and the proportion of primary fuels used for

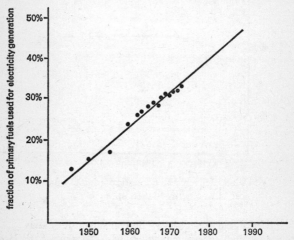

Figure 31  The fraction of primary fuel used to produce electricity.

generating electricity given by the trend in Figure 31. The result is shown in Figure 33. Since a substantial part of the growth in electricity usage is in furnace heating and domestic heating, both of which can be done as efficiently by a primary fuel, I will assume that *half* of the decline in efficiency of the fuel industries in the future is a real decrease in fuel-use efficiency. This has to be taken into account in calculating the future primary fuel demand of this scenario.

The calculation of future fuel demand is based on the assumption that G.N.P. continues to grow at an average 3 per cent per year.

*Futures I: Business-as-usual* **131**

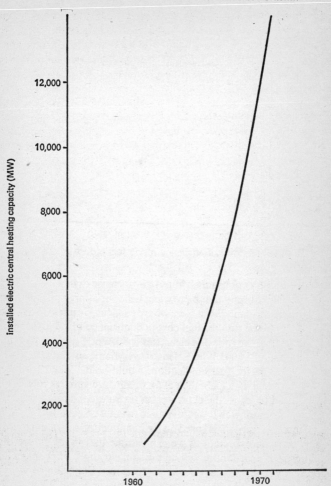

Figure 32 Exponential growth in electric central heating capacity.

This means that everyone has 3 per cent more real income each year; that is an increase over and above any changes in money values due to inflation. However, on its own this increase in

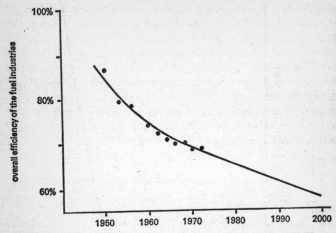

Figure 33  The overall efficiency of the fuel industries, 1950-72.

wealth doesn't tell us much about any increase in fuel demand. In Chapter 4 we found that there was a wide variation in the fuel cost of different commodities. Here we are particularly interested in the variation in the ratio of fuel cost to financial cost; kWht/£ value. For example the ratio for cigarettes is 8 kWht/£ and for oil (fuel oil) it is about 1,600 kWht/£. Thus if everyone spent their increased wealth on oil the fuel implications would be two hundred times greater than if they spent their extra wealth on cigarettes. So we not only need to know the extra wealth (extra expenditures) we also need to know how that wealth is used (what it is spent on).

Since this scenario is based on the continuation of present trends and personal expectations we can assume that in the future the patterns of expenditure will be the same as those of people who presently have a higher than average income. For example, if you presently earn £1,000, next year your real income is assumed to rise to £1,030. We further assume that you will spend your £1,030 in the same way as someone who presently earns £1,030. As it happens, there are lots of statistics available on the way that people with different incomes spend their money. A summary of these is shown in Figure 34. This shows that at low income levels a large proportion is spent on food, fuel and housing but at the higher income end a large proportion is spent on transport and services. When the data are examined in more detail it is found that there are other significant variations within each of these classes of expenditure. For example high-income groups spend a larger fraction of their food money on meat and a larger fraction of their transport money on private motoring than the low-income groups. Thus to relate these patterns of expenditure to energy demand it is necessary to break the expenditure down into many different classes, each with its own kWht/£ ratio.

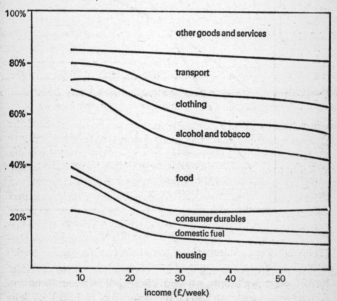

Figure 34 The changes in patterns of expenditure as a function of weekly Income. (Source: *Family Expenditure Survey*).

When this is done, the expenditure can be related directly to fuel demand. Results of such an analysis (Roberts, 1974) for transport, food and fuel (domestic use) are shown in Figure 35. This shows, for instance, that a household with an income of £10 per week (in 1968 £s) creates an annual fuel demand of $3.4 \times 10^3$ kWh

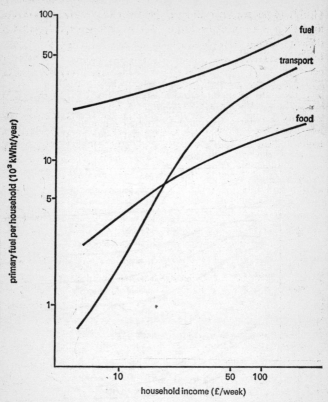

Figure 35 The fuel demand per household on a function of weekly income.

by its food expenditures, whereas a household with an income of £50 per week creates an annual fuel demand of $10 \cdot 1 \times 10^3$ kWh by its food expenditure. Thus these graphs provide a direct way of relating increases in income to increases in fuel demand.

There are four features of this method which need to be clarified. The first is that it presumes a constant population – in fact it assumes a constant number of households. The point is that the patterns of expenditure are characteristic of a household. Thus if the number of households were increasing at the same rate as

**G.N.P.** (3 per cent per annum) there would be no change in the average pattern of expenditure. This would give a different fuel demand from a calculation which assumes the number of households to be constant, so that each household income rises by 3 per cent and the pattern of expenditure changes accordingly. The second point is that these projections assume that the fuel-to-value (kWht/£) ratios stay the same. In fact we know that there are continuing improvements in technology, so that the kWht per item may be expected to decrease. However in my view these technological improvements are likely to be offset by declining

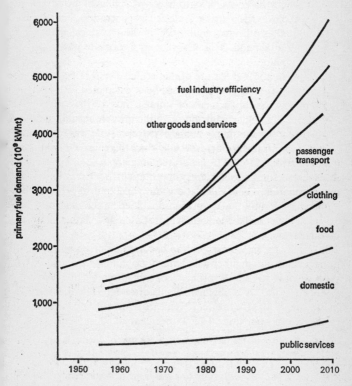

Figure 36  The total primary fuel demand for the b-a-u scenario.

grades (or increasing transport distances) of basic resources. For example a 10 per cent improvement in the efficiency of a steel furnace can be cancelled out by a decrease of 5 per cent in the average grade of iron ore. Thus without going into a lot more calculation about relative improvements and penalties we shall assume that the 1968 fuel/value ratios continue to be valid.

The third point is that the fuel costs deduced by these calculations are the primary fuel costs. As explained in Chapter 4, the fuel cost of a product includes an appropriate fraction of the fuels consumed by the fuel industries. However, earlier in this chapter we noted that there was likely to be a *real* decline in the efficiency of fuel use due to the increase in electricity generation. Thus I have incorporated an efficiency multiplier[3] in the overall calculation of primary fuel demand. This is shown as a separate part of total demand in Figure 36.

The fourth and last feature of these calculations is that the fuel demand of the transport sector has been calculated in a slightly different way. As can be seen in Figure 36 the growth in the transport sector is a major contributor to the growth in total fuel demand. I felt that the simple financial extrapolation was too crude an indicator for this sector, especially since there was a lot of other data available on future transport requirements. As it happens the transport sector is also extremely important in the evaluation of the other scenarios. These calculations have been put together in the Appendix and show that the projections based on personal mobility do give a lower fuel demand than those based on a simple financial extrapolation.

So far this scenario exploration has been a fairly sophisticated exercise in extending the fuel-demand graph up to the year 2010. Now we have to face the problem of trying to meet this demand.

The first strategy is obvious: simply extend the historical trends in fuel supply into the future. This means allowing the coal industry to decline to something like 75 per cent of its present output,

3. If the efficiency of the fuel industries in the year of interest is $\mu_t$ and the 1968 efficiency $\mu_{1968}$ then I have multiplied the calculated demand by a factor $(\mu_{1968}/\mu_t)$. This gives a fuel demand larger by a quantity $\Delta E$. I have assumed that half this increase is absorbed by a real increase in efficiency of use and the other half increases primary fuel demand.

allowing the production of North Sea gas to rise to its expected peak of 5–6,000 million cu. ft/day and taking up all the growth in fuel demand by increases in oil supply. This is shown in Figure 37 as the 'oil-option'. At the time of writing it appears a particularly foolhardy option, since the U.K. is still striving to cope with the

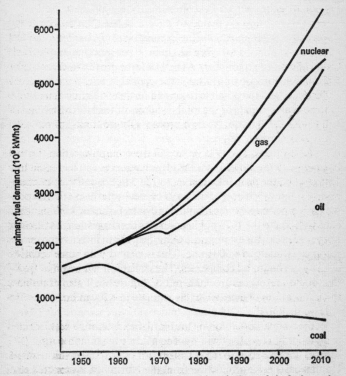

Figure 37 The oil-option, based on the historical trends in fuel supply.

balance-of-payments problems caused by its present level of oil consumption. However, some people are still taking this option seriously because of the promise of North Sea oil. Their optimism is based on a simple (and fallacious) calculation which suggests that since North Sea reserves are 3,000 million tons of oil and our present oil consumption is about 100 million tons/year the North Sea should see us O.K. for thirty years.

In 1974 the U.K. government published a booklet stating the position of North Sea oil (Dept of Energy, 1974). It took care to distinguish between *proven reserves* (which are quantities of oil known to exist and to be commercially exploitable), *probable reserves* (which are estimated to have a better than 50 per cent chance of being found and being exploitable) and *possible reserves* (which are estimated to have a less than 50 per cent chance of being found and being exploitable). At that time the proven reserves were 895 million tons of oil (m.t.o.), the probable reserves a further 1,095 m.t.o. and the possible reserves a further 960 m.t.o. The sum of these is an estimate of the total resources of the North Sea basin and equals 2,950 m.t.o. As time proceeds future 'finds' of oil will add to the proven reserves, but they will not change the total resource estimate. The 1975 version of the same publication (Dept of Energy, 1975) showed that the proven reserves had increased to 1,060 m.t.o., the probable reserves to 1,205 m.t.o. and the possible reserves had fallen to 835 m.t.o. The new total is 3,100 m.t.o., which is 5 per cent larger than the previous estimate. One significant addition to the 1975 publication was the suggestion that areas not yet licensed, for example the Celtic Sea, might increase the total resource estimate to 4,500 m.t.o. This is viewed with some scepticism by the major oil companies. Thus what follows assumes that the total oil resources are 3,000 m.t.o., but we shall also examine the implications of increasing this estimate to 4,500 m.t.o.

If we stick to the assumptions underlying this scenario, we have to let the companies exploiting the North Sea determine the rates of production. In fact they will want to achieve maximum output as soon as possible so as to repay their enormous capital investments. On this basis *peak* production from the North Sea is expected to occur in the 'early 1980s'. After that time production will fall. If these estimates are correct then the production of oil from the North Sea will follow the curve enclosing the shaded region in Figure 38. The total shaded area is equal to the total resource estimate of 3,000 m.t.o. The dashed production curve shows how the situation would change if the total resources were increased to 4,500 m.t.o. Also shown on Figure 38 is the demand for oil deduced

## Futures I: Business-as-usual 139

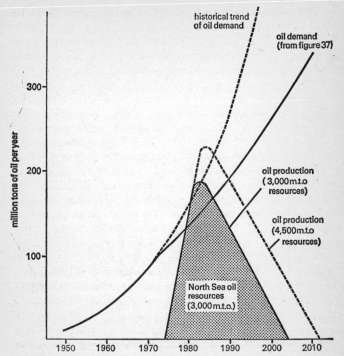

Figure 38 North Sea oil production assuming either 3,000 m.t.o. resources (shaded area) or 4,500 m.t.o. resources compared to the oil demand deduced from Figure 37.

from the 'oil-option' shown in Figure 37. This shows that if the total resources are 3,000 m.t.o. the supply of North Sea oil will only match demand from 1979 to 1986. If the total resources are 4,500 m.t.o. the period of self-sufficiency is extended up to 1991, an extra five years. However, more significant than the relatively short times of self-sufficiency are the effects that this venture has on the U.K.'s trade in oil. Figure 39 shows that the short period in which the U.K. is a net exporter of oil is followed by an astronomic rise in our demand for imported oil. It requires a benevolence beyond credibility to assume that ten years after ceasing to buy any oil someone in OPEC will be prepared to sell us more oil than we currently import. Furthermore the peak in world oil production is expected to occur in the period 1995–2000, so it is quite likely that not only will no one be *willing* to sell us any oil but that no one will have that much oil to sell. In summary North Sea oil does not justify a scenario based on continued growth in oil consumption.

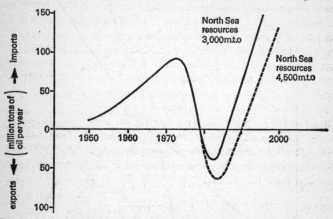

Figure 39   The U.K. trade in oil, 1950-2010, deduced from Figure 38.

This state of affairs is, I think, acknowledged by most numerate policy-makers. However, they do not simply give up the notion of continued growth in fuel supply. For the last twenty years the U.K. has been building up a nuclear-power capacity precisely to accommodate this anticipated shortage of oil. So now we have to investigate just how much nuclear power can reduce our demand for oil. Here there are two types of constraints, both of which have been discussed in Chapter 7. The first constraint is caused by the fuel investment problem and imposes a limit on the maximum rate of growth. The analysis in Chapter 7 showed that with present reactor designs and using relatively rich uranium ores the maximum rate of growth had a doubling time of about four years. With this rate of growth we found that half of the output of the stations finished was required for the construction of future nuclear stations, the other half contributing to the consumer demand for fuels. This is the rate of growth used in this scenario. It should be pointed out that if such a programme were undertaken the average grade of uranium ore would decline quite quickly. This would increase the fuel inputs needed by the programme and unless this increase could be offset by other technical improvements (for example by better enrichment technologies) the situation would be significantly worse than I have assumed.

The most limiting factor in nuclear power is the constraint not in the rate of growth but in the uranium resources. The energy analysis of mining and processing uranium in Chapter 7 showed that it is not energy-profitable to extract uranium from sources with a grade lower than 20 parts per million. This means that the traces of uranium found in all rocks and in sea-water[4] cannot be counted as a fuel resource, at least not for burner reactors.

The most optimistic estimates[5] of uranium resources that I have been able to find are shown in Table 14. As in the case of North Sea oil these estimates include a quantity of uranium which has not yet been discovered. The *known* reserves are less than half of the total shown in Table 14.

Table 14 Uranium resources (*outside communist countries*)

| ore cost $/lb. $U_3O_8$ (1973 prices) | concentration (ppm) | reasonably assured (million tonnes) | awaiting discovery (million tonnes) |
|---|---|---|---|
| 10 | 700–2,500 | 1·0 | 1·0 |
| 10–15 | 450–1,600 | 0·7 | 0·7 |
| 15–30 | 250–800 | 1·0 | 1·0 |
| 30–50 | 100–500 | 0·5 | 1·0 |
| | TOTALS | 3·2 | 3·7 |

4. The case against uranium extraction from sea-water is not the same as for rocks, which require energy for their crushing. In the case of sea-water the problem is one of processing huge quantities of water and, most importantly, keeping the depleted water separate from water yet to be treated. This water-handling problem could be overcome – but only by massive pumping schemes, which use too much fuel.

5. Not the largest guess, but the largest estimate based on some geological data and at least a plausible argument (Vaughan, 1974).

Now the U.K. would not have any trouble in meeting all its fuel demands if it could lay hands on all this uranium. However, an important assumption in this scenario is that all our major trading partners will also be following a b-a-u scenario, including, no doubt, growth in fuel consumption. They will run into precisely the same problem as the U.K. in finding a fuel substitute for oil, and have also been investing in nuclear power. Thus we have to calculate the U.K.'s share of these resources on the assumption that everyone else is doing much the same as the U.K. Very roughly it is assumed (by OECD for instance) that these resources will be divided three ways, between the U.S.A., Europe and 'the rest' (including Japan, Canada etc.).[6] Thus Europe's share of the total is 2·3 million tonnes. Of this the U.K. can probably claim about a fifth, or 460,000 tonnes of uranium, as its own. This places an upper limit on the capacity of nuclear burner reactors in the U.K.

The reactor assumed to supply most of the U.K.'s nuclear power is the SGHWR which, as we saw in Chapter 7, requires 3,743 tonnes of uranium to produce $115 \times 10^9$ kWh of electricity in its twenty-five-year lifetime. Assuming that the decrease in uranium ore grade (shown in Table 14) *is* offset by improvements in nuclear technology this implies that the 460,000 tonnes of uranium can produce $13,946 \times 10^9$ kWh of electricity. This total output will be spread over at least twenty-five years. In fact if we allow the nuclear capacity to grow at its maximum rate, we know that the production of nuclear electricity must follow the production curve shown in Figure 40. The flat part of the curve, from about 1995 to about 2020, represents the period when all the stations built up to 1995 are allowed to run their useful lifetimes. The total area under this production curve must equal the electricity produced from our quota of 460,000 tonnes, that is to say the area must equal $13,496 \times 10^9$ kWh. If the peak output is denoted by $E_p$ (kWh/annum), then the area under the production curve is approximately $37·5 \, E_p$. This means that $E_p$ is equal to $372 \times 10^9$ kWh/year. Each 1,000-MW station produces an output of $4·6 \times 10^9$ kWh/year, which implies a maximum installed capacity of 80,000 MW. This

6. This does not seem to me a 'fair' allocation of resources. It is however the allocation presumed by advocates of a b-a-u energy policy.

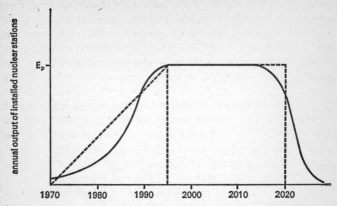

Figure 40  The output of nuclear electric stations as a function of time.

was the peak capacity used in the calculations in Chapter 7, so the results of that calculation (shown in Figure 25, p. 109) can be incorporated into this analysis. Figure 41 shows how the introduction of this nuclear capacity contributes to the supply of primary fuel. Also shown is the 'extra' primary fuel input needed by the rapidly growing nuclear programme. In Figure 41, the nuclear growth is not stopped abruptly (as in Figure 37) but slowed gradually after about 1985. Thus after 1985 most of the nuclear capacity (and not just half of it) contributes to primary fuel demand.

These calculations show that conventional nuclear burner reactors cannot do much to relieve the pressure on oil supplies. However the normal response to this is to point to the development of breeder reactors as a way of obtaining a lot more useful fuel from the same quantity of uranium. In fact the breeder reactor is able to use that part of natural uranium ($U^{238}$) which is not suitable for burner reactors. Thus the uranium limit does not impose any foreseeable constraint on the development of a breeder-reactor programme. The only problem is that we don't, as yet, have a commercially viable breeder reactor.

In fact the development of breeder reactors has been continuing for the past ten years and both the U.S.A. and U.K. are now at the 'prototype' stage. This means that if everything goes well we

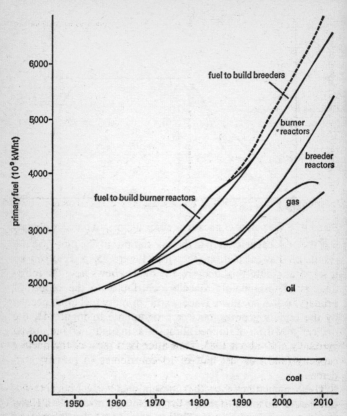

Figure 41  The nuclear supply option.

should have a full-size commercial breeder reactor by about 1985. If that proves to be a success, and if it does not require too many exotic materials, it could start to make a contribution to fuel supplies. However, like the growth in burner reactors, it will be limited in the maximum rate of growth which gives a real fuel saving. There is not yet sufficient data to do a thorough analysis of breeder reactors, but preliminary calculations suggest it should have an energy ratio at least twice that of burner reactors, so the number of breeders could double every three years and still show a net fuel saving. This rate of growth is, however, likely to be constrained by the availability of plutonium. Although the breeder reactor can breed its own fuel from $U^{238}$, the initial core of a breeder reactor has to be a plutonium core, and plutonium can only be produced in burner reactors. Unfortunately the breeder reactor produces enough plutonium to start another reactor only after twenty years' operation (that is to say the breeding time is twenty-years). Thus for the period 1985 to at least 2005 the number of breeder reactors will be limited by the plutonium produced by the limited stock of burner reactors. Thus, as shown in Figure 41, breeders only start to make a significant impact on oil supplies *after* 2005.

My conclusion from all this is that even with the fastest plausible development of nuclear power it cannot make a significant impact on oil demand. The oil demand in this nuclear option is shown in Figure 42, and you can see that if the total resources of the North Sea are 3,000 m.t.o. it has only extended the period of self-sufficiency up to 1989. Furthermore after that time it produces a similar rise in the demand for imported oil similar to that shown in Figure 39.

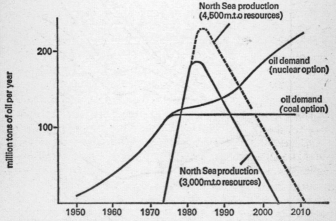

Figure 42  North Sea oil production compared with the oil demands deduced from the nuclear option (Fig. 41) and the coal option (Fig. 43).

## 146  Fuel's Paradise

Faced with this fairly dismal prognosis the policy-maker has to look to some other fuel source to reduce oil demand. North Sea gas is no good, since it is expected to reach peak production by 1985 and decline after that. I have already pointed to all the technical reasons why a *new* energy source (such as solar or wave power) cannot make a substantial impact in a growth scenario. So the only remaining alternative is coal, the industry that has been deliberately run down over the past twenty years.

A revival of the coal industry could, if coupled with the maxi-

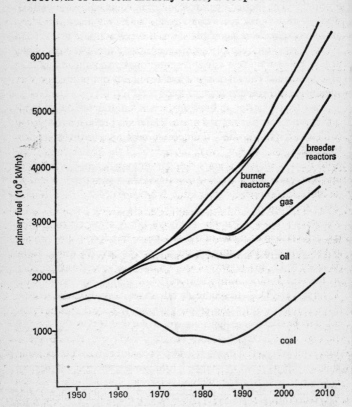

Figure 43  The coal supply option.

mum nuclear programme, reduce the demand for oil to a manageable problem. On paper the revival does not look too difficult. Figure 43 shows that it is only necessary to bring coal production back to its 1955 production by the year 2000 for the demand for oil to be kept fairly constant. If this is done, then, as is shown in Figure 42, the North Sea oil supplies will last until 1992. If the total oil resources are as high as 4,500 m.t.o., supplies could be extended to the year 2000.

Furthermore, if we can arrange either to constrain production to our own needs, or to organize a deal in which our exports of oil in 1979/92 are matched by imports in 1992/2005, then North Sea resources of 3,000 m.t.o. could see us through to the year 2005 and 4,500 m.t.o. could last up to the year 2020. Thus we have finally found the mix of fuels which can meet our demand curve, at least up to the year 2005 (when breeder reactors start to make a very significant contribution).

However, to stop our analysis here is to ignore all the serious problems that this scenario poses. There are problems associated with North Sea oil, with the development of crash nuclear programmes and, not least, with the revival of the coal industry. We shall discuss all these in some detail in a moment, but one must immediately emphasize the fact that this growth scenario has pushed all these fuel sources to their limit. If *any one* of them fails to meet its proportion of demand, there is no substitute available. This means that this is a 'fragile' scenario. If any of the presumptions are not verified or if any of the problems described below are not solved, the growth cannot continue – the scenario collapses. There is, as far as I know, no way of ascribing probabilities to events such as the formation of a successful cartel of copper exporters, or the failure of the U.K. to secure its fair share of uranium resources. These are essentially unpredictable events – just as it is not possible to predict where lightning will strike next. However, the fact that we can't predict lightning strikes doesn't stop them happening. Similarly our inability to put a probability on a future event doesn't mean it won't happen (unknown probabilities do not equal zero). This is the dilemma of policy formulation – all the important factors are unknown. All that is available is historical precedent and one's own intuition.

I have already made it clear that this scenario rests on the presumption that all the other industrialized nations will continue a b-a-u policy. If they follow some other policy it may make it easier to obtain uranium for example, but the change of policy will also remove one of our major export markets and thereby threaten our trading ability. Thus a change of policy in the U.S.A., Japan or even South Africa could put this scenario in jeopardy. Even more important is the presumption that the underdeveloped nations will not try to use their resources to shift the terms of world trade.

Leaving the validity of these presumptions for the moment let's examine the assumption made in regard to the North Sea oil resources. Here I implicitly assumed that the U.K. would retain all the oil in the U.K. sector for its own use. This is a very dubious assumption. Firstly if the U.K. remains a member of the E.E.C. then it has to face the fact that E.E.C. regulations forbid any member country from curtailing exports to other E.E.C. nations. France has already used this regulation to force the Dutch to sell them North Sea gas from the Dutch sector. Secondly the development of the North Sea has only been possible because of inputs of capital and expertise from the U.S.A., steel pipes from Japan and exploration and production rigs made in France. These countries view these contributions to North Sea development as giving them some claim to the oil found there. Certainly if the disposal of North Sea oil is left in the hands of the oil companies, as it should be in a b-a-u scenario, it will be sold to those prepared to buy it. Finally there is an incredible smugness associated with the belief that North Sea oil will see the U.K. all right when all around it are collapsing from oil shortages. France, Japan, West Germany and the U.S.A. have not got a North Sea – so their fuel problems are going to be a lot more acute than ours. It is naïve to imagine that their problems will not rub off on the U.K. To point to one simple connection, our major export to these nations is motor cars. We need to sell motor cars abroad to be able to afford to import food. How do you think you would fare as a car salesman if you were trying to sell a car to a Frenchman but wouldn't also sell him the petrol to drive it with?

The only way that the U.K. can keep all the North Sea oil to itself is for the rest of the world to find another, much larger, and hitherto unknown oil deposit. This would have to be about as large as the Saudi-Arabian deposit to alleviate the world-wide pressure on oil supplies. After seventy years of exploration this is a very remote possibility.

In discussing the development of nuclear power we considered only the resources of uranium and net-energy balance as constraints. In the U.S.A. the major constraint on the growth of nuclear power is not uranium availability, nor net energy considerations; it is pressure from environmental groups concerned with the safety of nuclear stations. In the U.K. there is some concern over the safety factors, but nothing like the organized legal pressure groups of the U.S.A. I suspect that a major reason why this development has not yet occurred in the U.K. is because so far only a few nuclear stations have been built near centres of population. Thus nuclear safety is not of concern to many people in the U.K. However the programme of burner reactors proposed in this scenario requires the construction of 80,000 MW of nuclear power – equivalent to forty giant 2,000-MW stations. Trying to find this many sites is no small task, and as more remote coastal sites are used up the pressures to build inland will grow, and so will the environmental pressure groups.

I do not have the space or the expert knowledge to make any contribution to the never-ending debates about nuclear safety. There are however two points pertinent to our evaluation of this scenario. The first concerns statements made by proponents of nuclear power that catastrophic accidents are 'impossible' or 'just cannot happen'. Such statements are logically nonsense. By definition no one can put a probability on an 'accident'. It is of course possible to put a probability on technical failures and I feel sure that nuclear engineers will make these probabilities suitably low (although they can *never* make them zero). But accidents always involve people. All the major aircraft crashes in the past five years have involved human fallibility; a pilot reading an instrument wrongly; a porter not shutting an aircraft door securely; a co-pilot forgetting to adjust engine speed at a critical time. These are unpredictable events, and if and when the first nuclear catastrophe occurs it will be in a way totally unforeseen by nuclear engineers and will almost certainly involve an error of human judgement.[7] So either people who declare that 'accidents are impossible' are fools, or they don't understand the word 'accident' – or they are simply con-men.

The second point is related to this discussion of accidents and is concerned with the consequences of a nuclear accident. I do not mean the consequences in terms of human lives or in terms of uninhabitable regions or in terms of radioactive releases. I am referring to the consequences for the nuclear-power programme. The debate about nuclear safety has become so polarized and the claims of nuclear proponents so absurd that one major accident will destroy the credibility of nuclear advocates. To the general public it will appear that the nuclear engineers were totally wrong (it will not appear that they were merely overstating their case). Thus the ferocity and polarization of the safety debate means that any nuclear accident will put the whole programme in serious jeopardy. (Let me put it this way. After all you have heard about the safety of nuclear reactors how would *you* react if someone wanted to build one ten miles from *your* house just after a reactor in France had blown up?)

The success of the nuclear-power programme in restraining oil demand beyond 1995 depended upon the successful development of breeder reactors. This is a fairly large stumbling block because it does require the engineers to solve difficult technical problems. Again I do not have the expert knowledge to comment on the probability of success or failure – it is simply another area of doubt. Allied to this are more serious problems concerned with the safety of breeder reactors. The reactor itself is more difficult to control, because of much higher energy densities, and the breeder fuel, plutonium, is far more toxic than other radioactive materials. Thus the safety question mark is even more serious for the breeder reactor than for the burner reactors.

The final area of problems is in the assumed revival of the coal industry, which was necessary to keep oil demand within the

7. This nearly happened in April 1975 when a nuclear reactor was put out of action by a maintenance worker burning through electric wiring with a *candle*. Did the designers of a nuclear station ever consider the possibility of someone wandering around the station with a candle!?

bounds of foreseeable supplies. This revival does not pose any serious problems from the point of view of coal resources, but there are problems involved in the resources needed to extract coal – namely coal-miners. One of the major factors leading to the historical decline of the coal industry was that the only way for coal to remain an economically competitive fuel was to improve the productivity of coal-miners drastically. This could only be done by closing down pits which, by their nature, could not be mechanized. This policy has been successful in that the output per man in the coal industry has risen from 294 tons/man/year in 1950 to 472 tons/man/year in 1973. However, a revival of the coal industry will reverse this trend because some of the old pits will have to be reopened. (The increase in production cannot be accomplished by opening up new pits because of the time needed to open a new pit and the limited output of any one pit.) This will have two effects. Firstly it will mean that the price of coal will have to rise relative to other fuels, owing to the decrease in productivity (output/man/year). Secondly it will significantly increase the number of coal-miners needed in the U.K. To attract more men to the industry will require increased wages, which will further push up the price of coal. The total number of miners needed by the year 2000 is shown in Table 15 to be no less than 540,000, which is 300,000 more than the present workforce! And this is needed after a twenty-year policy of running the industry down and encouraging men to move out of mining villages. I, for one, have no desire to become a coal-miner – and I seriously doubt whether there are 300,000 men in the U.K. willing to take up mining as a career. Perhaps there would be if you paid them £5,000 a year (1973 prices) – but that would put the price of coal at about twice the present oil price (per kWh). Rises in the price of fuel are a strong inflationary factor, since a fuel price rise puts up the price of every other commodity. (Witness the inflation in 1974/5 due to an oil price rise.) Thus there seems no way of reviving the coal industry without removing one of the preconditions needed for growth – namely no inflation and 'business confidence'.

Before leaving this discussion of the b-a-u scenario there is one other aspect which, for me, is perhaps the most important argument against this policy option. Earlier we saw that, while it was not

## 152  Fuel's Paradise

Table 15  Output and employment in the coal industry

| Year | Output/man/year (tons coal) | Total output (million tons) | Employment (thousand men) |
|---|---|---|---|
| Historical | | | |
| 1950 | 294 | 202 | 687 |
| 1960 | 305 | 196 | 642 |
| 1970 | 459 | 154 | 335 |
| 1973 | 472 | 120 | 254 |
| Needed | | | |
| 1980 | 400* | 155 | 387 |
| 1990 | 460* | 210 | 456 |
| 2000 | 500 | 270 | 540 |

* initial fall due to reopening old pits; subsequent rise due to dramatic improvements in coal-mining technology.

possible to extend any detailed analysis of scenarios much more than thirty years into the future, it was necessary to cast a longer-term look to make sure that the direction of development makes sense. The glance cast in the longer term, represented by the simple extrapolation shown in Figure 44, shows that just beyond the time period of our analysis the U.K.'s fuel consumption curve hits the 'thermal limit' discussed in detail in Chapter 6.

The limit is set between $7,500 \times 10^9$ kWh and $10,000 \times 10^9$ kWh, corresponding to three to four times our present fuel consumption. This level of fuel consumption occurs sometime between 2020 and 2030. As indicated earlier, I think that if we do not constrain our fuel consumption before this limit is reached we could face some serious climatic changes, especially down-wind from urban-industrial areas (which by that time would form a solid belt across England from Liverpool to Dover).

At the moment the principal reason for continuing growth is that it conforms to everyone's personal expectations of a better tomorrow and that reducing growth would cause serious social problems. (More on this in the next two chapters.) How much more difficult will it be to slow growth by the year 2000? At that

## Futures I: Business-as-usual  153

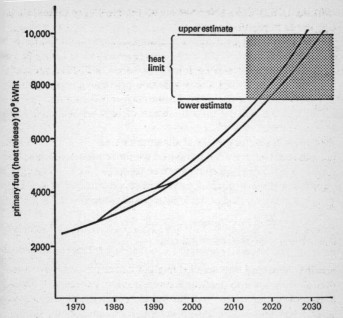

Figure 44  The b-a-u fuel demand and the estimated heat limit for the U.K.

time we will have experienced fifty years of non-stop growth and the social repercussions of changing will be much greater than those prevailing today. Thus in the year 2000 it will be more difficult to design a fuel policy that avoids the heat limit than it is today.

Later we shall be looking at the relative advantages and disadvantages of all the scenarios, so let me summarize the main conclusions of our exploration of b-a-u.

The scenario has the advantage that it conforms to the social expectations built up over the past twenty-five years of continuing growth. It is actively supported by all the important political institutions in the U.K. The scenario presumes that:

(i) the U.K.'s trading partners will adopt b-a-u policies;
(ii) the underdeveloped countries will not form cartels or otherwise change the terms of world trade;

## 154  Fuel's Paradise

(iii) the U.K. will be able to retain all the North Sea oil and still trade in motor cars, etc.;
(iv) there will not be a major nuclear accident within the time-period;
(v) sufficient sites will be found for forty 2,000-MW nuclear stations and that these will not provoke environmental pressure groups able to delay the nuclear programme;
(vi) the U.K. will be able to secure at least 460,000 tons of uranium;
(vii) breeder reactors will be available after 1985;
(viii) the coal industry can be revived without causing inflation;
(ix) 300,000 extra coal-miners can be found;
(x) either the heat limit doesn't exist or growth can be slowed sufficiently quickly after 2000 to avoid exceeding it.

Finally it should be emphasized that in this scenario there is only *one* energy policy option and that this involves obtaining maximum production from all the available fuel resources. This can only be accomplished if *all* the presumptions listed above are verified.

# 11 Futures II: Technical-fix

As with the business-as-usual scenario our first task is to understand the basic philosophy behind a technical-fix type of energy policy and establish the reasons for choosing this policy option. The basic reason for choosing this option is that it appears to offer much greater future *security* than a b-a-u policy and at the same time does not seriously upset social expectations or life-styles. In this policy option the problems of finding future fuel supplies are significantly reduced by reducing the demand for fuel. However, it is presumed that the reduction in fuel demand can be accomplished only by measures which do not significantly alter present life-styles or decrease the standard of living. Essentially someone opting for this type of policy is saying that the probability of failure of a b-a-u policy is so large that they would prefer to opt for a planned reduction in the growth of fuel consumption rather than have it forced on them.

If at the same time as reducing fuel consumption there is to be a continued improvement in the standard of living and continued satisfaction of personal expectations, there must be fairly substantial improvements in technical 'efficiency'. Efficiency in this context refers to the ratio of fuel use to the provision of goods and services, housing and personal mobility. To accomplish this requires the introduction of a slightly new definition of 'better'. The point is that in the past 'better' housing has meant rebuilding old or derelict houses so as to provide more space and more services to those living in the house. In this scenario 'better' housing also means houses with more insulation, so that the same level of warmth and comfort can be achieved with a smaller fuel consumption. Similarly the present notion of a 'better' car seems to roughly correspond with a bigger, more expensive model; in this scenario a 'better' car would be one with similar levels of passenger space and comfort but a much smaller fuel consumption.

This does not mean that in some sense energy should become an important factor in 'value' or 'welfare'. It seems to me quite arbitrary that at the moment the 'best' cars are simply the biggest. This value judgement on cars has been deliberately encouraged by the advertising techniques used to sell cars. I see no reason why in the future car manufacturers should not try to sell their 'better' cars on the basis of their fuel consumption. Indeed this is already happening. From January to October 1974 I collected all the car advertisements that I happened to see in newspapers and magazines, and every single advertisement mentioned the fuel consumption of the car. This shift in the general perception of what constitutes a 'better' car has come about because of the rise in price of petrol. Similarly I expect that in the near future the increased costs of home heating will cause house buyers to pay more attention to its insulation characteristics and less attention to the glamorous 'picture-windows'. So the shift towards a technical-fix philosophy has already begun. However, to have a significant impact on fuel demands, this policy will have to be supported by deliberate government policies. For example the rate at which the stock of existing houses could become better insulated could be greatly accelerated by a government subsidy for house insulation. Trends towards smaller motor cars could similarly be encouraged by fiscal policies such as increased fuel taxes and a sliding scale of car road-fund licence.

Where fuel consumption represents a direct personal expenditure, that is to say where it is a perceived cost, then these simple price changes may be enough to reduce fuel demand significantly. However, as we saw in Chapter 4, at least three quarters of our fuel demand is generated indirectly by our purchases of goods and services. Under these circumstances individuals are unlikely to express a strong enough preference for goods with smaller fuel costs to encourage manufacturers to change their methods of production. For example a loaf of bread produced from wheat grown using organic fertilizer involves 7·3 per cent less fuel consumption than one grown using inorganic fertilizers (see Figure

8, p. 54). For a 30p loaf of bread this fuel saving corresponds to about 0·3p. Thus if the government wants to see fuel conserved in the industrial sector it will have to persuade business by methods outside the market place.

To a large degree this is under government control. The government can encourage or discourage the use of fertilizers by its agricultural policy. By giving subsidies to local authorities, the government could encourage district heating schemes and the recycling of paper and metals in domestic refuse. However, perhaps the most important way for government to persuade industry to adopt a 'technical-fix' philosophy is for it to explain the basis of its fuel policy very carefully. If the government makes it clear that we cannot guarantee keeping all the North Sea oil, if it points to the problems of securing sufficient uranium supplies and if it points out the difficulties in increasing coal output without increasing coal prices then industry could be persuaded.

The rise in price of oil has already substantially altered government and industrial attitudes to fuel supplies. At one time policies were formulated on the basis of the 'best guess' of the future availability of fuel. Now there is a trend towards looking at the 'worst guess' also, and trying to take some steps which act as an insurance against the worst. The only type of insurance policy in fuel supplies is to have some 'slack' in the supply situation. While demand is pressing against the maximum possible supply, any technical or political hitch causes chaos. However, if there is the possibility of increasing supplies, then failures (such as a setback in the development of breeder reactors) or unforeseen political events (the U.S.A., Canada and Australia forming a uranium cartel) can be accommodated. Since the b-a-u policy fully stretched *all* the available fuel supplies the only way of building in this reserve ability is to reduce fuel demand.

Although it is implicit in this scenario that the fuel-conservation measures should have a minimum impact on life-style, there must be some impact. For us *all* to continue with the same expectations and life-styles we have to adopt a b-a-u policy. Thus the aim of this policy option is to minimize the impact on life-style while at the same time causing a sufficiently large reduction in fuel demand. These carefully chosen words conceal an extremely difficult problem of policy evaluation. How can you measure 'impact on lifestyle'? And what constitutes a 'sufficiently large reduction in fuel demand'? In theory such questions are resolved by economic analyses which try to evaluate 'costs' and 'benefits' and weigh them in financial terms. In practice it is nowhere near as easy as this. In the first place no two economists will agree on the precise ways in which future costs and benefits should be calculated. This is because there are implicit value judgements in these types of calculations on which people have different opinions.[1] Apart from anything else the value placed on a fuel saving must depend upon the probability of some fuel production process not succeeding – and we have already noted that such probabilities cannot be evaluated.

In practice policy-makers are more constrained by what they *can* do than by problems of deciding what to do. For example no cost-benefit analysis will cause a U.K. government to close down the U.K. car industry in the next thirty years. This is simply because the car manufacturers, road builders, associated trades unions etc. will be able to exert a strong enough political pressure to prevent any such government legislation. This type of constraint means that in practice a 'technical-fix' policy would be adopted wherever it was politically possible and where the social dislocation could be shown to be less serious than the implications of an equivalent fuel shortage. For example there seems to be a large measure of public support for passing legislation to outlaw disposable bottles. At worst this would require part of the present labour force which makes bottles to become transporters or packers of returned bottles.

In trying to put this scenario together I have chosen a number of fuel-conservation measures which seem to me to be both politically feasible and, in principle, economically justifiable. I have not the knowledge or data to do more than indicate the types of economic and social consequences of these measures. However, the techniques of energy analysis do enable us to evaluate fairly precisely the fuel savings which they imply. Thus this picture of a 'technical-

---

1. It is no good simply looking at the total monetary value of costs and benefits if they accrue to different people at different times. How do you balance a loss of income of £10 per week to a school-teacher against a future capital gain of £20 to a millionaire?

fix' scenario is lopsided in that it has a lot of detail on fuel implications but nowhere near enough on social and economic implications. The major fuel-conservation strategies used in this scenario are:

(i) reduce the growth of electricity generation so that the electricity generated does not exceed that which can be 'usefully' used. Also introduce financial penalties for the use of electricity for space heating, cease advertising of electrical heating appliances etc.;
(ii) reduce passenger transport fuel by a shift towards smaller cars and *reducing* the shift away from public transport;
(iii) reduce domestic fuel use by better insulation of houses and the introduction of a few combined heating schemes;
(iv) reduce the fuel used to produce commodities by obtaining industrial low-grade heat from rejected heat and by reducing the quantities of material per product;
(v) reduce the fuel used in producing food by reducing the *growth* in the use of inorganic fertilizers and food packaging;
(vi) reduce the fuel used in retail and distribution sectors by encouraging better insulation of commercial premises, recycling of paper, reduction of packaging;
(vii) reduce the fuel consumption of the public-service sector by improving the insulation of government offices.

These may sound fairly insignificant policies, but they are sufficient to hold U.K. fuel consumption below $3,000 \times 10^9$ kWh up to the year 2010. This figure was reached in 1990 in the b-a-u scenario; thus the policies have given us a twenty-year 'breathing' space. The reason why they have had a significant impact on total fuel demand is that I have chosen to reduce the *rate of growth* at precisely those points where *fuel consumption* was increasing most rapidly. At the same time many of these 'lever' points are not very significant from the point of view of life-style. I seriously doubt whether many people would feel 'deprived' or 'worse-off' if the food they purchased had just one layer of packaging instead of two. Similarly, unless you attach great value to living or working in glass houses, better insulation standards are unlikely to decrease your assessment of your standard of living.

The calculation of the fuel savings from these policies is fairly straightforward. For example the policy of reducing the proportion of primary fuels converted to electricity leads to a direct saving of the 'fuel industry inefficiency' shown in Figure 36 (p. 135). As explained below, the reduction in the use of electric central heating coupled with improved home insulation reverses the trend of increasing domestic fuel requirements shown in Figure 30 (p. 129). The factors which lead to significant savings in the transport sector are explained in detail in the Appendix. The only other policies which have to be evaluated in fuel-cost terms are those associated with food production and recycling.

Let us start by looking at these two policies, taking food production first. Table 16 gives a detailed breakdown of the items included in the 'production and provision' of food in 1968. Also shown is the ten-year growth of these items based on the business-as-usual scenario. Some of these growth rates can be estimated by using other data; in particular the use of fuels and fertilizers on the farm is well documented.[2] Other growth rates, such as the fuel and equipment used in 'cafés, restaurants, pubs, etc.', can be fairly well estimated because they correspond to a particular class of consumer expenditure (which, as explained in Chapter 10, can be used to calculate future expenditures). The other growth rates given are 'intelligent guesses', adjusted so that the weighted average growth rate agrees with that calculated from total expenditure on food. To estimate the corresponding growth rate in the technical-fix scenario I have to estimate the rates of growth of each of these items and calculate a new weighted average.

In this scenario the total food output is assumed to continue to increase by about 1 per cent per annum. In the past this growth has also been accompanied by much greater 'industrialization' of agriculture, in particular the use of inorganic fertilizers. This has increased the productivity per *man* in agriculture but not dramatically changed the productivity per *acre*. Since inorganic fertilizers have a high fuel cost (15–90,000 kWht/ton) they make a much larger contribution to the total fuel bill than to the total financial

2. However, I have not simply taken the historical growth rates, which are very large. Instead I have assumed that the use of fuels and fertilizers is saturating and used growth rates lower than the historical rates.

bill. Thus in this scenario the *growth* in inorganic fertilizer use is drastically cut from 15 per cent (over the ten-year period) to 5 per cent. The continued growth in food output is made up by increasing the use of organic fertilizers, which have no 'fuel cost'. The growth in use of fuel on the farm is also slightly reduced owing to a shift to smaller tractors and more economical use of all machinery. (Note that, because of higher fuel prices, this is less than in the b-a-u scenario, but there is still an increase from today's use of chemicals and tractors.)

Table 16  *Fuel consumption and growth rates in the food sector*

| | Fuel cost $10^9$ kWht | Per cent of total | Growth in ten-year period | |
|---|---|---|---|---|
| | | | b-a-u (%) | t-f (%) |
| Fuel used on farm | 34·3 | 8·4 | 20 | 15 |
| Fertilizers and chemicals | 46·1 | 11·3 | 15 | 5 |
| Freight transport | 13·0 | 3·2 | 10 | 10 |
| Food industry fuel | 66·1 | 16·2 | 10 | 5 |
| Packaging | 42·4 | 10·4 | 20 | 10 |
| Other industrial inputs | 55·5 | 13·6 | 15 | 20 |
| Food shops: | | | | |
| fuel | 64·5 | 15·9 | 35 | 15 |
| equipment | 11·0 | 2·7 | 20 | 20 |
| Cafés, restaurants, pubs: fuel | 53·0 | 13·0 | 35 | 20 |
| equipment | 21·0 | 5·2 | 20 | 25 |
| TOTALS | 406·9 | 100% | | |
| | Weighted average | | 21·1% | 13·5% |

The reduction in the *growth* of fuel use in shops, the food industry and 'cafés, pubs etc.' is brought about by straightforward technological improvements. All these sectors are large consumers of electricity, mostly for low-temperature heating purposes. It is assumed that the policy of reducing the growth in electricity supply will encourage these sectors of the food industry to obtain their low-temperature heat directly from other fuels. Notice that, if all the b-a-u growth were attributable to increasing use of electricity, this tactic would reduce the growth rate by a factor of 3. I have assumed a much smaller reduction in growth, which should mean that these sectors actually have available the same quantity of useful heat. Since this policy involves a change in technology it is necessary to allow a slightly higher growth rate in the purchase of equipment so as to accommodate the changeover. The net result of all these changes is that instead of a 21·1 per cent increase in the use of fuels by the food industry in ten years this scenario requires only a 13·5 per cent increase over the same period. The total fuel demand of the food sector in the two scenarios is shown in Figure 45.

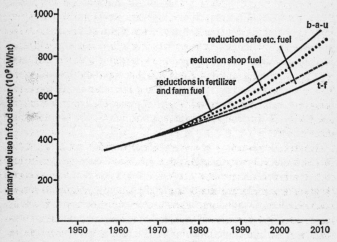

Figure 45   Primary fuel use for food in the b-a-u and t-f scenarios.

The use of electricity for providing low-temperature space heating is also reduced in the domestic, commercial and public-services sectors. As pointed out in Chapter 5, the use of electricity for space heating represents gross inefficiency. Exactly the same amount of heating can be provided by directly using a third of the quantity of fossil fuels needed for the electrical system. It is difficult to find reliable data on exactly how much electricity is used for space heating. The evidence presented in Figure 30 suggests that the increase in primary fuel requirements for domestic heating, cooking etc. is due to a shift to electricity, since the total heat delivered to households has not changed over the past ten to twenty years. One piece of fairly reliable data comes from the Electricity Council, and gives the total installed capacity of electric central-heating systems. Since electricity consumers have to notify the electricity boards of these types of installations, the data is likely to be accurate. It is also fairly astounding. The changes in installed capacity of electric heating is shown in Figure 32. It is an exponential growth with a doubling time of 2·6 years! Needless to say such fast growth cannot continue for long, but by 1973 the installed capacity of 25,000 MW would probably have consumed about $150 \times 10^9$ kWh of primary fuel.[3] This represents more than 5 per cent of total primary fuel consumption. Thus, if this trend were reversed and the same amount of heating achieved using fossil-fuel systems the total energy saving would be about 4 per cent of total primary fuel consumption. However, this only takes into account electric central-heating schemes. While these are likely to be the largest consumers, there will be a comparable consumption of electricity in ordinary electric heaters, convector heaters, fan heaters, etc.

From a simple energy-analysis point of view this extravagant use of electricity for space heating appears absurd. However, there are other aspects to the problem. In 1969 when I was choosing a space-heating system for a recently purchased house I chose to install electric storage radiators. The reasons for this were basically economic. The purchase of the new house had used up all my savings, so I had no capital available. The mortgage repayments on the house were already so high that I could not afford to increase my mortgage unless, by so doing, I made an equivalent monthly saving in running costs. At that time the cost per kWh for oil was about

3. This is based on an average load factor of 17 per cent. The load factor takes account of the fact that an off-peak system is only operating 8 hrs/day (33 per cent) and will only be used for about half the year. Then the total primary fuel consumed is given by

$$4 \times (25,000,000 \text{ kW}) \times (8,760 \text{ hrs/yr}) \times 0.17 = 148.92 \times 10^9 \text{ kWh}.$$

0·18p/kWh. For electricity the cost was about 0·95p/kWh – but off-peak electricity cost only 0·33p/kWh. The capital cost of an oil- (or gas-) fired central-heating system at that time was about £1,500; the cost of the off-peak electric system was £150. Assuming difference in running costs was £60/year, which would require that I would consume about 40,000 kWh/year on heating, the twenty-three years of operation to make up the difference in capital costs. So I chose the electrical system; and, as can be seen in Figure 32, so did many other people!

The reason why the economic argument gives a completely different answer from the fuel-efficiency argument hinges around price of off-peak electricity. Had I been able to purchase only normal tariff electricity (at 0·95p/kWh), the difference in running costs would have been £300/year and it would have required only four years' operation to make up the difference in capital cost. So I would have chosen the oil system. The reduction in price on the off-peak electricity tariff arises because the CEGB (and other generating authorities) want to try to smooth out the very large fluctuations in the demand for electricity. From the CEGB's point of view about half the cost of generating a kWh of electricity is due to fuel costs; the other half arises because of interest payments on the capital invested in the power station. Clearly, the more electricity produced by the power station, the smaller will be the capital cost per kWh. The fuel costs per kWh are relatively fixed, so the more electricity that is generated the smaller are the unit costs. Thus the trick of making money in the electricity industry is to keep as many power stations as possible operating at maximum capacity. This poses a problem for the CEGB, since it is legally bound to install sufficient generating capacity to meet *peak* demand, not average demand. Thus the only strategy open to it is to try to make the average demand as close as possible to the peak demand which is why a reduced tariff is offered to off-peak users.

From a fuel-efficiency point of view it is also sensible to try to make average demand equal to peak demand. This is because it requires a lot of fuel to start up a generating set, and even if they are not switched off the fuel efficiencies of generators decrease when they are run at lower power. However, the crux of the issue is that by asking the CEGB to make its own operations as efficient (in financial terms) as possible the overall system efficiency is reduced. If the extra costs to the nation due to the use of electricity for space heating were *chargeable* to the CEGB, it would have to take the use of electricity (and not just its production) into account and would have to change its policies.

To some degree this has happened already. The rise in price of oil (and coal) in 1974 significantly shifted the ratio of capital costs/fuel costs and made it comparatively less important for the CEGB to sell off-peak electricity. I have also noticed the differences in our heating bills. In fact my heating bills are now two or three times greater than my capital investment in 1969 – so I (and I guess many other people) am preparing to change my heating system. One of the advantages claimed for electricity is that it is a more flexible fuel and can be controlled a lot easier than other fuels. However these arguments do not apply to electric off-peak systems. Such systems are very *inflexible*. I estimate that this lack of flexibility means that I actually purchase 30–40 per cent more heat than I need.[4] Thus the case against electric central-heating systems is overwhelming.

In this scenario I have assumed that the government removes the off-peak tariff and takes legislative steps to reduce the sales of electrical heating appliances. For wholesale measures such as this to be politically acceptable the government may have to consider subsidizing domestic and commercial users of electric space heating so as to allow them to change over to a fossil-fuel system. The shift away from electric space heating is estimated to reduce the growth in the fuel demand of goods and services by 0·5 per cent per year (falling to 0·25 per cent reduction per year by 2010); to reduce the growth in 'public services' by about 0·1 per cent per year and to reduce domestic fuel growth by about 0·5 per cent per year (falling to a 0·2 per cent reduction by 2010).

These reductions in the rate of growth maintain the same output of goods and services, but at a slightly better overall fuel efficiency. In the domestic sector further significant fuel savings are achieved by improving house insulation. A programme of increasing house

4. There is an alternative, which is to keep the heaters on a lower setting, but then I would get 20–30 per cent less heat than needed on the cold days!

insulation initially increases the fuel consumption of the U.K. because of the extra production of materials (glass, glass-fibre etc.). However, the insulation measures envisaged in this scenario pay for themselves in fuel terms within two or three years.[5]

As an example let's calculate the fuel costs and savings of double glazing a standard three bedroom semi-detached house. The house has about 300 square feet of window area which has to be double-glazed. The ex-factory fuel cost of window glass is 5·58 kWht/sq. ft. However, by the time it has been transported to the retailer and a proportion wasted in being cut to size the final product has a fuel cost of about 9·5 kWht/sq. ft. A 6×4 ft window has an area of 25 sq. ft and needs about 25 lb. (11·3 kg) of aluminium framing and about 1 lb. of rubber or putty sealing compound. These correspond to 0·45 kg aluminium/sq. ft and 0·04 kg rubber/square ft. Table 17 shows that the total fuel cost of all the materials is about 47 kWht/sq. ft window, so our house needs 14,000 kWht of fuel for double glazing.

*Table 17  The fuel costs of double glazing*

|  | kWht/sq. ft |
|---|---|
| Glass | 9·51 |
| Aluminium* (0·45 kg at 80 kWht/kg) | 36·29 |
| Rubber (0·04 kg at 45 kWht/kg) | 1·63 |
| TOTAL | 47·0 |

* assuming that all this extra aluminium has to be found by increasing primary production.

To calculate the saving in fuel used to heat the house we need to compare the rate of loss of heat through single- and double-glazed windows. Table 18 gives the standard 'U-coefficients' (or heat loss coefficients) for various parts of domestic buildings. If we assume that the U-coefficients for windows are slightly reduced by curtains, the single-glazed windows lose 5W/m²/°C and the double-glazed

5. Clearly at some point the fuel cost of an extra layer of insulation will not be worthwhile. However, the average U.K. house is so badly insulated that it will take a long time to reach such a situation.

*Table 18 Standard U-values for external walls and windows*

|   | U-value (W/m²/°C) |
|---|---|
| 1. Solid brick wall plus 16-mm plaster | 3·0 |
| 2. Cavity wall plus 16-mm plaster | 1·5 |
| 3. Cavity wall plus cavity filling and 16-mm plaster | 0·7 |
| 4. Single-glazed windows | 5·6 |
| 5. Double-glazed windows | 3·2 |
| 6. Pitched roof with roofing felt and plaster board on ceiling joists | 1·9 |
| 7. As 6, with 50-mm glass fibre | 0·51 |

windows lose 3W/m²/°C. For average home temperatures the product of temperature (°C) and time (hours) is about 53,000 °C hours/year, so the total fuel saving in the home is

(change in U value) × (area windows) × (°C hours/year)

$$= 2·0 \times 300 \text{ sq. ft} \times \frac{1}{10·76} \text{ (m}^2\text{/sq. ft)} \times 53,000 \text{ (Wh/year)}$$

$$= 2,955,390 \text{ Wh/year}$$

$$= 2,955·4 \text{ kWh/year}$$

This is the saving in heat delivered to the house. To find the saving in primary fuel we have to take into account the efficiency of fuel use in the house (boiler efficiency) and the efficiency of the appropriate fuel industry. For example if the house is oil-heated then, taking account of the time when the boiler is running at less than peak demand, we might expect an average boiler efficiency of 60 per cent. The efficiency of the oil industry is about 90 per cent (see Chapter 4), so the primary fuel needed to supply the 2,955 kWh of heat is

$$2,955 \times \frac{1}{0·6} \times \frac{1}{0·9} = 5,473 \text{ kWht.}$$

Since the fuel cost of double glazing was found to be 14,000 kWht, this indicates a 'pay-back' time of 2·6 years. Just as in the analysis of nuclear power stations (see Chapter 7), before this information can be used as a basis for determining policy it is essential to look at the time variation in fuel consumption associated with the double glazing of one house. This shows that the *first* result of a double-glazing policy is to *increase* fuel consumption. At a national level the increase in fuel consumption and its duration are determined by the rate at which houses are converted to double glazing. In practice any such policy will be constrained by how many people can be persuaded to install double glazing and by how fast the glass and aluminium industries can increase their output. Of these the glass industry is likely to be the most limiting, since if *all* new houses, offices and shops plus some fraction of existing buildings are to be double-glazed then the output of the sheet-glass industry will have to be more than double. (It would have to almost double simply to allow double glazing of all new houses.) This would take at least four to five years to accomplish. If by the year 2010 we want a large proportion of all the housing stock (say 80–90 per cent) to be double-glazed, the number of old houses converted will have to be very large. At the moment the U.K. builds about 300,000 new houses each year and has a total stock of about $20 \times 10^6$ houses. To convert all the existing houses in the thirty years 1980–2010 requires 600,000 to be converted each year. Clearly this will not happen immediately. Figure 47 shows a plausible rate of converting the existing stock of houses to double glazing. Combining this graph with the fuel costs and savings of one house, shown in Figure 46, gives the total fuel-demand curve shown in Figure 48. The total saving in domestic fuel demand, assuming that there is no net increase in total housing stock[6] is $97 \times 10^9$ kWht by 2010. This is equivalent to 4·2 per cent of 1970 fuel consumption and, spread over the forty years up to 2010, represents an annual *reduction* in domestic fuel *growth* of 0·1 per cent.

Similar calculations on roof insulation and cavity-wall insulation[7] give slightly higher energy savings with a corresponding larger reduction in domestic fuel growth. In the b-a-u scenario the growth in domestic fuel demand was made up by growth in the sectors

---

6. As in the b-a-u scenario I am assuming that the total number of households remains constant. Hence the number of houses is roughly constant.
7. It is assumed that by the time this policy is fully implemented 75 per cent of houses will have cavity walls, since by 2010 new building will have replaced more than half the existing housing stock.

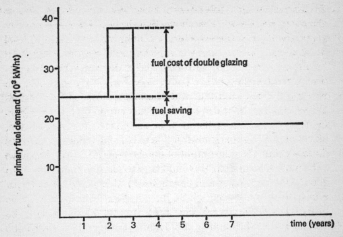

Figure 46  Fuel costs and savings associated with double glazing a house.

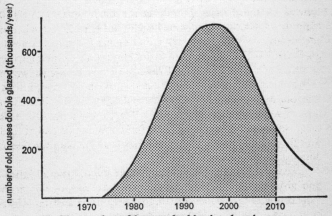

Figure 47  The number of houses double-glazed each year.

shown in Table 19. Although the direct fuel use is the slowest-growing sector it dominates the growth because it accounts for 83 per cent of the total fuel demand. The four energy-conservation measures adopted in this scenario – reduction of electrical heating,

# 170 Fuel's Paradise

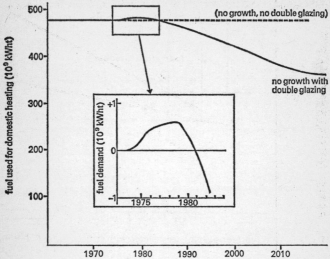

Figure 48 The fuel required for domestic space heating assuming that houses are double glazed at the rate shown in Figure 47.

Table 19 Growth of the domestic sector

|  | 1968 | | Growth in ten-year period (per cent) | |
| --- | --- | --- | --- | --- |
|  | Fuel $10^9$ kWht | per cent of total | b-a-u | t-f |
| Direct fuel use | 676 | 83 | 13·0 | 5·0 |
| House construction | 33 | 4 | 28·0 | 28·0 |
| Furniture and consumer durables | 75 | 9·3 | 29·0 | 20·0 |
| House and home retail outlets | 30 | 3·7 | 23·0 | 15·0 |
|  | 814 | 100 |  |  |
|  | weighted average growth | | 15·4 | 7·6 |

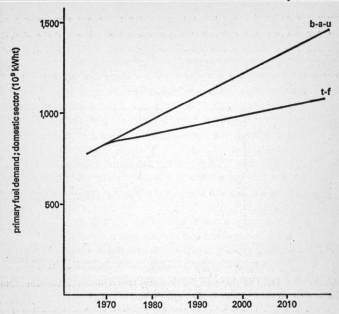

Figure 49 The total domestic fuel demand in the b-a-u and t-f scenarios.

double glazing, roof and wall insulation – together reduce the growth in direct fuel consumption to 0·5 per cent per annum. The reduction in growth of the other items in Table 19 is due to improvements in methods of production and the reduced use of electricity in shops, showrooms etc. These measures reduce the growth of the domestic sector to 7·6 per cent in a ten-year period, with the resulting savings shown in Figure 49.

The last of the policy measures mentioned earlier in the chapter was the encouragement of recycling materials. This has been suggested as a way of reducing fuel consumption, since it requires a lot less fuel to produce materials from scrap than from ores (Chapman, 1974(e), 1975). However, recycling is unlikely to reduce the consumption fuel in the U.K. This is because for most materials the U.K. has three sources of supply, namely domestic primary production (from ores), secondary production (from scrap) and imports. An increase in secondary production from scrap materials is most likely to reduce U.K. imports, not U.K. primary production. Thus an increase in recycling may well do a lot to help the balance of payments, but may slightly increase U.K. fuel consumption, not decrease it. A preferable strategy is to promote the re-use of commodities. The distinction between re-use and recycling is that, whereas the latter reclaims materials which can subsequently be reprocessed, re-use actually makes use of a product without any significant reprocessing. Perhaps the best illustration of this is the glass milk bottle. The re-use of glass milk bottles *as* glass milk bottles (and not as a source of glass) represents a significant fuel saving (Boustead, 1974). If all glass containers were re-used in a similar way, the fuel saving could be as high as 1 per cent of the total fuel demand. This is a useful once-and-for-all saving, but hardly significant in the evaluation of energy policy.

The net effect of all the policies outlined earlier is summarized in Figure 50, which shows the total fuel demand and its division into the six sectors of consumption. This shows that the policies have had the desired effect in significantly slowing down the rate of growth, so that by the year 2010 the total primary fuel demand is some $2,300 \times 10^9$ kWht less than in the b-a-u case. This reduction in primary fuel demand, equivalent to the total fuel demand in 1968, has the effect of taking the pressure off all the fuel resources. This, in its turn, opens up a much wider range of options of possible fuel supplies and, more importantly, leaves sufficient flexibility for unforeseen problems or shortages to be avoided. The scenario thus becomes a more 'robust' policy option. The point is that in the b-a-u scenario there was only *one* possible mix of fuel supplies which satisfied the total demand, and this involved pressing every fuel supply to maximum production. In contrast this scenario allows a choice between different mixes of fuel supply.

The first supply option, shown in Figure 51, is a 'nuclear option' which pushes ahead as fast as possible on burner-reactor development and also makes use of breeder reactors from 1985 onwards. The coal industry is allowed to continue its decline; the gas industry reaches peak production around 1985 and has faded by the year 2010. This combination of policies means that there is a significant and sustained reduction in oil consumption from 1980

*Futures II: Technical-fix* 173

onwards. As you might expect this puts North Sea oil resources in quite a different light.

One of the assumptions made in establishing this scenario is that the government would be prepared to take the formation of energy policies seriously. The implementation of the policies set out earlier involves considerable government intervention in private industry. To be consistent with this we should also allow the government to intervene in the production of oil from the

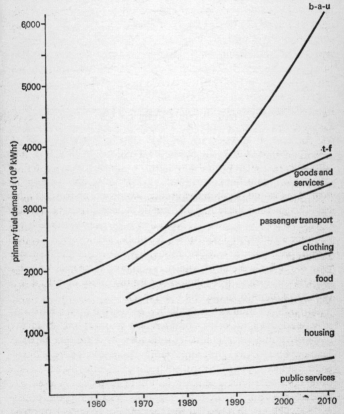

Figure 50 The total primary fuel demand for the t-f scenario.

## 174 Fuel's Paradise

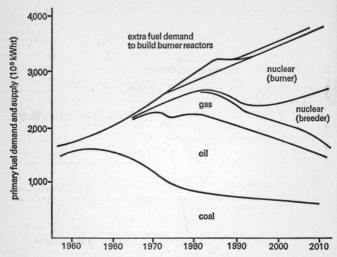

Figure 51 A nuclear supply option for the t-f scenario.

North Sea. A fairly moderate intervention (which was also implicitly needed in the b-a-u scenario) would be to constrain peak production from the North Sea to 120 million tonnes of oil per year. This would result in the production profile shown in Figure 52. Also shown is the falling oil demand deduced from the 'nuclear option' shown in Figure 51. Note that now, even with the lower resource estimate of 3,000 m.t.o., North Sea oil lasts into the next century and that thereafter the quantity needed is much smaller than in the b-a-u scenario. This may still pose some problems, but only half as serious as in the b-a-u case (60 m.t.o. in 2010 as opposed to 120 m.t.o. in the best b-a-u case).

As an illustration of the flexibility introduced by the reduction in overall demand let us consider how a policy-maker in this scenario could accommodate a failure to introduce breeder reactors by 1985. It is not important whether the failure is due to technical factors, to successful environmental lobbying or to some other considerations. The important thing is to be able to accommodate the shortfall in total supplies. Presumably the policy-maker will know that breeder reactors will not be appearing on his

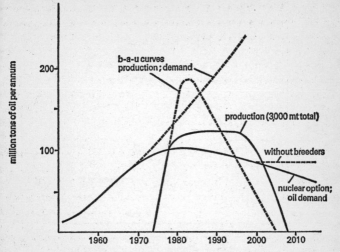

Figure 52 North Sea oil production, based on 3,000 m.t.o. resources, and the oil demand in the nuclear supply option.

supply graphs in the time period 1980–85. As an immediate stop-gap measure he could probably allow the demand for oil to increase, thereby taking the oil-demand curve closer to the production curve in Figure 52. However, this leaves a fairly serious problem beyond the year 2000. To avoid this all he has to do is immediately to stop the decline in the coal industry and to take steps to enable it to increase output from the year 2000 onwards. The supply option is shown in Figure 53 and the resulting oil-demand curve is shown dashed on Figure 52. The moderate revival in coal output shown in Figure 53 does not pose any of the serious problems of the coal revival in the b-a-u case. Firstly there is a ten-year period, 1985–95, when output has only to be held constant. This of course raises the option of trying to introduce new 'conservation policies' in that time period so as to make up for the lack of breeder reactors. After 1995, assuming that no conservation policies are implemented and that the North Sea production curve has not been further 'flattened', coal output has to increase by 10 million tons by 2000 and by 60 million tons by the year 2010. Assuming the same coal-

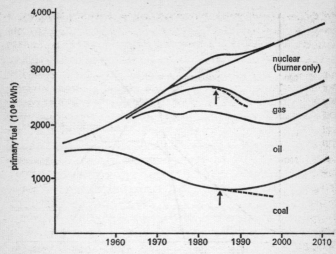

Figure 53 The change of supply option needed if breeder reactors are not available by 1985. The arrows show when the supply mix has to change from that shown in Figure 51.

miner productivities set out in Table 15 (p. 152) for the b-a-u scenario, these increases in production require an additional 20,000 miners by 2000 and a further 40,000 miners by 2010. Alternatively assuming that the coal workforce stays at its 1974 level of 250,000 men, the 1980–95 productivity of 440 tons/man/year has to be increased to 480 tons/man/year in 2000 and 680 tons/man/year in 2010. These calculations show that there is no serious difficulty in making up for the absence of breeder reactors and that there are at least two policy options available for dealing with this problem.

Although this supply option coincides with present plans for the supply of primary fuels to the U.K. it may not be an appropriate supply option for this scenario. The reason for this is that present technology can only utilize the heat generated in a nuclear reactor to produce electricity. It is therefore not sensible to obtain 50 per cent of the primary fuel input from nuclear power (by 2010) in a scenario which has drastically reduced electricity demand as a way of conserving fuel. Furthermore the trend shown in Figure 51 is for

*Futures II: Technical-fix* 177

an even larger fraction of primary fuel to come from nuclear sources. This difficulty could be avoided if techniques were developed for using nuclear heat directly. Such developments are currently being discussed, which means that by 2010 they *could* be developed. But if they are not developed then the supply mix shown in Figure 51 will put in jeopardy the main fuel-conservation policy of this scenario.

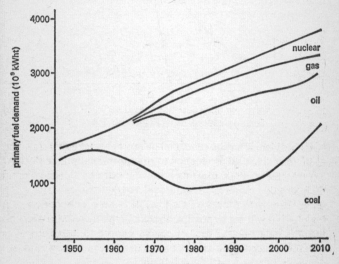

Figure 54  A non-nuclear supply option for the t-f scenario.

To avoid this problem, and to further emphasize the range of options available in this scenario, I have constructed a non-nuclear supply option. This could also be chosen as a supply policy in the event of a public rejection of nuclear power following some accident or dramatic increase in costs (as discussed in Chapter 10). The supply option is shown in Figure 54 and is based on a policy of meeting the growth in fuel demand by increasing the use of oil and coal. In fact I have assumed that the policy-maker adjusts the demand for oil so that it follows the North Sea production curve more closely, as shown in Figure 55. To avoid a serious oil im-

178 *Fuel's Paradise*

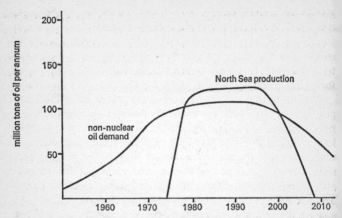

Figure 55 North Sea oil production (based on 3,000 m.t.o. resources) and the oil demand deduced from Figure 54.

ports problem after the year 2000 he must also take steps (around 1980) to enable coal production to rise substantially after 1995. At a productivity of 500 tons/man/year the extra coal production in the year 2000 requires an additional 100,000 miners – still much fewer than *needed* for the only feasible b-a-u supply option. If it looks as if it will not be possible to achieve this increase in coal output, the policy-maker can, of course, choose to introduce some new fuel-conservation measure so as to reduce the growth in fuel demand after the year 2000 still further.

It is important to realize that this extra policy option, of introducing some new conservation measure, is not available in the b-a-u scenario, but poses no serious difficulties in a t-f scenario. As explained in Chapter 10, the b-a-u scenario is based on a conventional growth philosophy. This makes it very difficult for the government to implement a conservation measure suddenly. However, the t-f scenario is based on the presumption that the government is successful in engendering shifts in value judgements, so that, for example, 'better cars' are the more fuel-efficient. With this basic shift in approach and philosophy already accomplished an extension to another area of fuel consumption is quiet feasible. The reduction of fuel demand also has the important consequence that it gives policy-makers more time to choose and implement their policies. In the b-a-u scenario the pressure on every single source of supply meant that any failure produced a very quick shortfall of supplies. In this scenario we have found that the policy-maker has ten to thirty years to implement some alternative, depending upon the type of 'failure' involved.

With all these advantages it may strike you as absurd that any government should choose to follow a b-a-u policy and not a t-f policy. This is because we have yet to face the most serious problems which implementing a t-f policy poses. What I have shown is that a few carefully chosen fuel-conservation measures can give a significant reduction in total fuel demand, so that all the resource and technological problems are reduced to a manageable level. There is sufficient slack in the demands imposed on different fuel sources to make it possible to change the primary mix of fuels smoothly and without hiccups over periods of ten years or more. Some of the policies do not reduce our material standard of living. For example total personal mobility (passenger transport) is the same as in the b-a-u case. Also the growth in food output and in building and heating houses is the same as in the b-a-u case. In all these areas fuel demand has been reduced by making the appropriate technology more fuel-efficient.

This leads directly to the two major problem areas associated with this scenario. The first is that of actually accomplishing the improvements in efficiency. Although this raises no technical problems, it does raise the problem of how to persuade individuals and institutions to accept new philosophies, new value judgements etc. The second difficulty with this scenario is that technical efficiency cannot continue to be improved indefinitely. In a system such as that prevailing in 1975 it is easy to improve efficiency so as to bring about significant fuel savings. However, in the year 2005 it will be a lot more difficult, since most of the obvious waste will have already been swept aside. If during these thirty years people's expectations have continued to rise, it is likely that from that time onwards the increase in fuel supplies will revert to the b-a-u growth case. This is shown in Figure 56. In effect the technical-fix policy has given us twenty-five years' breathing space. The fuel problems of 1995 become the fuel problems of 2020. Only when we get to 2020

180   *Fuel's Paradise*

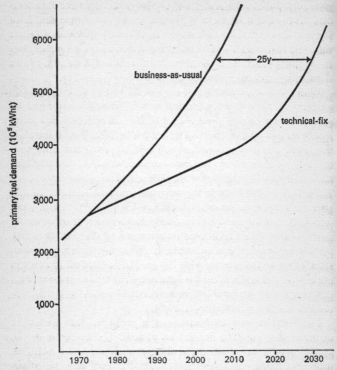

Figure 56 On a longer time-scale the t-f scenario may have the effect of buying 25 years, after which demand rises again at the b-a-u rate.

in the t-f scenario we have lost two important policy options – namely North Sea oil and a t-f option.

Now it may be that all we can ask of a policy adopted now (1975) is that it removes a foreseeable problem (in 1995) far enough into the future that it need not bother us. I have already pointed out that the range of technological options by the year 2020 *may* be *much* larger than those which we could expect by the year 1995. However, it is still disturbing to find that all we have done is shift a basic problem from this generation on to the next.

But this isn't the major problem area. The real problem area is that associated with persuading individuals and institutions to accept new policies, new philosophies etc. The government policies needed to bring about the reduction of fuel demand are listed fairly glibly earlier in this chapter. Let us now take a more realistic (pessimistic?) look at these.

The first, and by far the most important, was the reduction in growth of the electricity industry. Coupled to this was legislation which made it either impossible or prohibitively expensive for people to use electricity for space heating. The technical arguments are overwhelming, but politically this cuts very little ice. The electricity industry has experienced twenty-five years of steady 7 per cent growth (doubling time: ten years). It will not look kindly on a reduction to 2–3 per cent growth for the next twenty-five years. Furthermore the CEGB has become a very powerful political lobby. This is associated mostly with its control over investment policy. In 1970 the CEGB invested £486m, equivalent to £1·3m every day. This represented 15 per cent of all capital investment by manufacturing industries in 1970 and is by far the largest single source of investment in the U.K. Now a reduction of growth of the electricity industry from 7 per cent per annum to say 3 per cent per annum would reduce the investment needed over the next five years from something like £2,500m to about £750m. The reduction is larger than the reduction in growth because the CEGB's investments in 1970 were designed to meet loads expected in 1977. The change in growth of electricity from 1970 would mean that the demand expected in 1977 would not arise until after 1980. The reduction in investment, from £2,500 million to £750 million, poses very severe problems for the heavy capital industries, particularly metal manufacture and electrical engineering. To put it crudely it means that a heavy electrical engineering firm can expect its orders to be reduced by a factor of 3 or 4, which either means it goes out of business or has to find some other way of using its equipment and labour force.[8] The electrical engineering trade will also suffer severe reductions in orders owing to the allied policy of reducing

---

8. The capital formation industries are always the worst hit by reductions since their output is tied to growth rate, not to total production. In this case the time lags of the electricity industry exacerbate an already serious problem.

the sales (and hence production) of electric space-heating devices.

What all this means is that there will be very strong political pressures which will try to persuade the government *not* to cut back on electricity production. The pressures will take all forms. There will undoubtedly be a massive advertising campaign by the electrical trades to point out the higher fuel efficiency of electricity for some applications. There have already been highly misleading advertisements from the Electricity Council claiming that electric heating is more efficient than any other. (This is, of course, based on the quantity of heat delivered to premises and does not take into account the losses in converting a primary fuel to electricity.) There will be extravagant claims about the amount of unemployment which would be caused by the policy, and there will be lobbying on the old-boy network. The problems of following a b-a-u scenario will be dismissed, ignored or obscured by irrelevant arguments about technicalities. In short it will require an extremely determined and very strong government to enforce such a policy.

Similar remarks apply to other policies which have industrial implications. Car manufacturers will raise hell at the suggestion of a motor taxation policy that attempts to price 2,000-cc. cars out of the market. The large chemical industries will produce all types of arguments to prevent a reduction in the growth of their fertilizer divisions. The arguments are all too easy to see. Car manufacturers will point to their export earnings on large cars and claim that they need a strong home market to be able to continue in the same way. Fertilizer manufacturers will raise the spectre of food shortages or, even worse, play upon urban ignorance of agriculture by pointing to the undesirability of growing potatoes on fields covered with cow-shit. The packaging industry (the manufacturers of plastics for cartons, paper for boxes etc.) will point to all the advantages of 'handi-pacs' and their ability to make food more hygienic.

In all these cases policies of reducing *growth* will be misrepresented by claims that they are reducing *total production* and hence taking us 'backwards'. I fear that these misrepresentations will be successful precisely because all our expectations are based on continued growth, not just on continued satisfaction of our *present* demands.

It may appear that I am painting a picture of irresponsible

business-ogres eager to ensure the continued growth of their own profits regardless of the costs to society. This is a naïve and mistaken view of the situation. All the people opposing technical-fix policies will be doing so because they are convinced that their particular activity is contributing to human welfare. For example most of those working in the electricity or fertilizer industries will be impressed by the arguments in support of electricity and inorganic fertilizers. If they aren't, then the chances are that they will either have moved out of those industries or at least will not have attained positions of power and influence in them. Which company would promote a manager who insisted on questioning whether the company's product was really worthwhile? The individuals who succeed in a company are precisely those who can argue best for the continued use of its products in ever-increasing volume. These are also the individuals who will perceive policies of reduced growth as constituting the most serious threat. While an institution is growing, individuals can always hope for promotions or other increases in job satisfaction as a consequence of institutional growth. A static institution can only promote its staff at the rate at which directors die. There is thus a strong dynamic in all institutions to grow so as to enhance all its employees prospects (see Galbraith, 1972). A government policy to reduce this growth will be perceived as a direct threat to these better prospects. Under these conditions arguments will be misunderstood and very rapidly reduced to emotive (irrational) levels.

So, in summary, I would say that although a technical-fix scenario will not run into any serious technical/resource problems it will run into very serious social/political problems. The source of political opposition will be those bright and dedicated men and women who have made particular industries their careers. There will not be any conscious conspiracy nor any conscious misleading of the public by these people. However, they will act together and present distorted arguments against government action. It would perhaps be possible for one government to undertake reducing the growth of the electricity industry during its five years in office. For a government to take on the electrical industry, the fertilizer industry, the car-manufacturing industry and the packaging industry at the same time spells suicide. The uproar would be almost as damaging as a political revolution, with loss of business confidence, drastic reduction in investment and consequent unemployment. The government would then be defeated and replaced with one committed to 'restoring the economy' – that is to say putting us back on a b-a-u policy path.

By now you are doubtless thinking, 'Well, if this is the case why on earth have we spent the last chapter going through detailed calculations on an impossible scenario?' There are two simple answers to this. As pointed out in Chapter 9, I do not think there is any 'problem-free' future. To achieve any kind of future requires us to solve some problems. If you think that the t-f scenario is attractive from a resource point of view you should *not* overlook the serious social/political problems it brings in its wake. Perhaps by pointing out to you how the institutions involved will react to t-f policies you will not be misled by their arguments.

The second answer is that there is quite a lot of evidence available now (early 1975) to suggest that the U.K. government is actually heading towards a type of technical-fix policy. For example, the price of electricity has been pushed up and the government has announced that the off-peak tariff will be a lot more expensive from April 1976. To a large degree the trend towards a t-f policy has been brought about by the change in the relative price of fuel since 1973. In effect all the government is doing is pushing a little bit more than the market forces would do alone. Thus they are not exposing themselves as significant policy-makers. For example, the significant rise in price of electricity had already, by January 1975, caused the demand for electricity to fall below its 1970 level and caused the CEGB to reduce its investments drastically. Because this arose through the market mechanism it has not raised howls of protest, but it *is* obvious to most observers that the government could have subsidized the CEGB and kept the demand up. It is encouraging that they didn't. However, if ever it becomes necessary for the government to appear as a significant policy-maker, the political furore described earlier will become apparent.

Summarizing the main features of this scenario we have seen that its advantages are that it:

(i) conforms to personal expectations of increasing material possessions and comforts;
(ii) is a robust energy policy, since there are several possible mixes of fuel supplies which can meet demand;
(iii) avoids most of the technical/resource problems of the business-as-usual scenario;

The main disadvantages of the scenario are that it:

(iv) still relies upon the ability of the U.K. to trade with the rest of the world on the present basis (the reduction in oil imports has made the trade situation slightly easier, but has not substantially altered the U.K.'s dependence on others for food, fuel and raw materials);
(v) will run into severe political opposition from those institutions whose activities are curbed so as to reduce fuel demand. The opposition could prevent any government from introducing all the necessary measures.

## 12  Futures III: Low-growth

**15 October 1976:** Members of CIPEC, the cartel of copper exporters, today finally managed to reach agreement with the major bauxite producers on pricing and output policies. With the exception of Australia, all the countries involved agreed to increase the price of copper and aluminium ores by 50 per cent. They also agreed to peg total output to January 1976 levels until they meet again, which will not be until January 1977. The implications of these agreements for the U.K. are an additional £400m import bill and serious problems for the engineering trade.

**1 February 1977:** Sweden today announced its support for developing nations attempting to reduce international inequality. It announced a 50 per cent rise in the price of iron-ore exports and pledged that the extra revenue would be placed in UNAID accounts for development projects. This decision adds another £50m to the U.K. import bill and poses headaches for the B.S.C.

**12 June 1977:** The melt-down of the 600-MW PWR outside Lyons last week has effectively stopped the French nuclear-power programme. The total radioactive release was much less than previously feared because the upper containment structures survived the worst of the accident. The calm weather also did much to reduce the scale of the disaster. The accident has had repercussions in the nuclear programmes in U.S.A. and West Germany as well as the U.K.

**July 1977:** The Congressional committee investigating the safeguarding of plutonium since April 1974 has presented its report to the President. It is believed that the report strongly recommends that the development of breeder reactors should be shelved for at least ten years while new safety systems are developed.

October 1977: The urban guerrilla movement in New York succeeded in hijacking 20 kg of plutonium en route from a reprocessing plant. The President has placed an immediate and total ban on all plutonium shipments and has instructed the FEA and USAEC to shelve plans for breeder reactors 'for at least ten years – perhaps indefinitely'.

January 1978: The white regime in Rhodesia finally collapsed. The incoming black nationalist administration immediately banned all exports of chromium to 'unfriendly' nations and cut off all trading relations with the U.K.

March 1978: Canada, Australia and the U.S.A. published a joint statement on their agreed policies to impose quotas on uranium exports. The countries denied that they were forming a cartel and insisted that the measures were designed to protect their own indigenous fuel resources at a time when other sources of fuel were not available to them. The measures mean that supplies of uranium to the U.K. will be cut by a factor of five.

April 1978: Argentina today let it be known that in the future it would sell beef only to nations prepared to sell fuel in return. This brings the international trade war to a new pitch and is likely to set a precedent for many other primary producers.

June 1979: The latest trade figures show that the U.K.'s exports have slumped to the lowest level since 1950. A government spokesman said that there were three basic causes; the non-existent foreign car markets; the sharp decline in purchases of capital equipment of all types; and the success of the Japanese in capturing no less than 30 per cent of the total world market of manufactured goods.

July 1979: The Minister of Trade and Industry today disclosed plans to increase oil production from the North Sea to enable the U.K. to make up its trade deficit by sales of oil.

May 1980: The Scottish Nationalists today claimed responsibility for blowing up the pipes bringing oil on-shore from the Piper and Forties fields. Oil companies estimated that it would take at least six months to replace the broken pipes at a total cost of £200m. Meanwhile the U.K. will have to manage with half the oil supply it needs. A government spokesman said that stocks would be enough to last for only two months.

In theory, making decisions should be reducible to a straightforward logical argument. To illustrate this let us assume that you have changed your job and want to choose between two identical houses. Let's suppose that house A is nearer to your new job than house B, but that house B has a much larger garden. To resolve your choice of house you have to estimate the relative costs and benefits of each house. Since you want to compare the totals for each house you have to use some common factor. It must clearly be money. Now let's suppose that after a lot of thinking and arithmetic you can put monetary values on such things as transport and growing your own vegetables in each of the two houses. Let's also assume that you consider two kinds of future, one in which the price of transport is very high and the other in which the price of food is very high. Then you could arrive at the simple decision-matrix shown below. Here a benefit is represented by a positive (+) value, a net cost by a negative (−) value.

|                              | House A | House B |
|------------------------------|---------|---------|
| Future I: expensive transport | +£150   | − £70   |
| Future II: expensive food     | − £50   | +£200   |

Which house do you choose? According to decision-making theory you can make a rational choice only when you have ascribed a probability[1] to each type of future. Let's assume that you assume

---

1. I am using probability in its strict mathematical sense. Under these conditions an event with a probability of 1·0 is certain and an event with a probability of 0·0 certainly never occurs. Since the example considers only two types of future the sum of the probabilities of each must equal 1·0.

both futures are equally likely and that there is no other possibility. Then each future has a probability equal to 0·5. Now your choice can be resolved by multiplying all the costs and benefits by their appropriate probability and working out the total for each house. For house A the total equals $\{(150\times0\cdot5)+(-50\times0\cdot5)\}$, which is +£50. For house B the total equals $\{(-70\times0\cdot5)+(200\times0\cdot5)\}$ = £65. Thus you should choose house B, since this choice most probably gives the largest benefit.

The point of this simple example is to point out that according to this theory your choice depends critically upon the probabilities you attribute to each type of future. For example if you think that future I is a lot more likely than future II, so that future I has a probability of 0·9 and future II a probability of 0·1, then the totals are +£130 for house A and —£43 for house B. Thus with these new probabilities you should choose house A. This type of reversal due to a change in probabilities is quite normal and emphasizes the importance of assessing the probabilities of different types of future in coming to any kind of decision. However we already know that there is no way of estimating the probabilities of the types of events which shape the future. This is not merely to say that it is difficult, but that there is no logical way of measuring or calculating the probabilities involved.[2] It was to emphasize this point that this chapter began by describing a sequence of disastrous, but plausible, events. I don't think these events *will* come to pass, nor do I think they *will not*. I think that anyone will have to admit these are possibilities which cannot be predicted. This means that policy-makers cannot justify using methods such as those described earlier because they do not know and *cannot* know any of the probabilities involved. They just have to live with the fact that some or all of the events described earlier may occur.

There are other reasons for rejecting this formal approach to making decisions, all of which are important in establishing the basis of the scenario described in this chapter. The next most

---

2. Normally probabilities can either be measured, for example, by determining how often a tossed coin comes down heads, or can be calculated from arguments which rely on deterministic knowledge of the events, such as the fact that a coin has two sides and is symmetrical. Neither of these procedures can be applied to events which are significant in formulating policy.

obvious problem concerning the theoretical approach described is that there is no value-free way of attributing costs and benefits to policies, or of comparing costs and benefits of one policy with another. There are two aspects to this argument. The first is concerned with the use of prices of goods and services as a way of estimating their *value*. This may be an acceptable procedure for *existing* conditions, but cannot be defended when applied to policies which set out to change the conditions. Furthermore there is no logical way of estimating the value of 'goods' which are not incorporated in financial transactions. How much is clean air worth to you? How much would you be prepared to pay to avoid a specified level of radiation from radioactive materials? If *you* are the only person involved in the consequences of a particular decision, you may be able to guestimate your own values of such 'goods', but without a market in 'clean air' there is no way to judge other people's valuations. The second aspect of the procedure which raises serious problems is associated with the comparison of costs and benefits when they apply to different people at different times. Earlier I posed the problem in terms of a policy which resulted in a benefit to a school-teacher of £10 now *and* a benefit to a millionaire of £20 in five years' time. How can these benefits be compared? Here there is no need to argue about the technicalities of different procedures, merely to emphasize that *any* procedure involves making a value judgement. Traditionally economists have taken monetary values at face value and ignored any discrepancies in income. They assumed that this was a 'value-free' judgement, whereas in fact they have made a very strong value judgement in favour of the existing distribution of income.

This last argument leads to the most important, but least tangible, argument against this type of formal decision-making procedure. It is also the point at which advocates of the low-growth scenario depart from the conventional philosophy of making decisions. The point is that choosing a policy is not just a problem in finding the optimum allocation of resources over time. All policy decisions are political directives for the future and will reflect the type of future that the policy-maker *wants* to see. An individual's perception of the world, including his hopes for the future, will colour his judgements in all sorts of ways. At a fairly trivial (yet not unimportant!) level an individual's political affiliation will strongly influence how he attributes values to various effects of policies. However, the most fundamental level at which differences in 'world view' have effect is not in arriving at different answers to the same questions but in the types of questions which are asked.

This is an important point which is all too often overlooked. For example the differences in approach and values of Conservative and Labour governments may arrive at different answers to questions such as 'at what rate should the coal industry be run down?' or 'which type of nuclear reactor is best for the U.K.?' However, neither party questions whether the coal industry *should* be run down or whether the U.K. should have *any* nuclear reactors. Such questions, which are never discussed or debated, are the questions which both a Conservative and a Labour 'world view' answer in the same way. However, it is precisely these questions which are most influential in determining the future in the sense that by not discussing them whole areas of policy options are not even considered. Advocates of a low-growth scenario start by insisting that these questions should be the ones under discussion, and they arrive at unconventional answers.

This brings me to the reasons advanced for choosing a low-growth scenario as a basis for formulating energy policy. There are two justifications offered for this type of future. The first is that the direction of our development is wrong because it is taking us further away from a stable or sustainable way of life. For example it is pointed out that if growth continues then sooner or later we will run into the heat limit, or run out of water or run out of land or some other resource. The alternative to the growth direction is to try to organize our productive activities so that they are sustainable, which means we should start to move towards a long-term equilibrium with our environment. For instance this would require a decision to try to reduce our dependence on stocks of non-renewable fuels and to start using the renewable sources such as solar, geothermal and wind power. In essence the argument is saying that it is foolish to live off one's capital (stocks of fossil and nuclear fuels) when one has available a perfectly good income (renewable fuel sources).

The second justification for this scenario is based on a value judgement that survival is more important than increased affluence. Thus those advocating this scenario concentrate on the ability of the U.K. to survive the worst kind of future. In effect this discounts any benefits from policies which increase the risk of non-survival and attaches greatest value to policies which incorporate some form of insurance against disaster. This argument is often presented in silly terms which, for example, equate a reduction of total food supply with mass starvation. This ignores the fundamental ability of all humans to adapt to very severe circumstances and still 'survive'. The vast majority of the population of the U.K. would physically survive, albeit with great discomfort, if the U.K. were suddenly unable to import any food or fuel. What is at stake is not physical survival but the survival of our social structure and the comforts which we have all come to expect. Thus the implication is that the U.K. should make itself as independent as possible of the types of events described at the beginning of this chapter. This suggests that we should try to develop towards a state of self-sufficiency and is quite contrary to the conventional economic philosophy of maximizing benefit by increasing trade. The idea of self-sufficiency is, however, consistent with the long-term goal of moving towards some type of equilibrium with the environment.

There are other reasons why this type of scenario is seriously considered by some people. For example they argue that factors such as 'loss of community', 'lack of job satisfaction' and 'loss of amenity' are more important than the benefits from increased material possessions. To a degree this reflects the relative affluence of those advocating the scenario, but the change of philosophy is consistent with the removal of the basic physical deprivation in our society. However, the purpose of this chapter is not to argue for or against this particular scenario. I have spent some time explaining the basis of it only because it is not normally included amongst options for the future. Now I want to examine its implications as ruthlessly as for the other scenarios and point out its advantages and the problems it raises.

The energy policy normally associated with a low-growth philosophy is one in which all the technical improvements possible are made, and in addition some policies are aimed at changing life-styles so as to achieve further fuel savings. Briefly the major policies which I have taken to put this scenario together are

  (i) a reduction in total passenger transport coupled with a significant shift towards public transport;
 (ii) all the technical improvements in passenger transport used in the t-f scenario;
(iii) the conversion of most houses in the U.K. to either some form of solar heating or district heating (using the output of electricity generating stations);
 (iv) a reduction in the use of electricity for space heating as in the t-f scenario;
  (v) the use of alcohol derived from straw to provide fuel for use on farms;
 (vi) a net reduction of the use of fertilizers and packaging (as opposed to a reduction in growth rate in the t-f scenario);
(vii) a reduction in the growth rates in the clothing, goods and public-services sectors in accord with the overall philosophy of stabilization.

The net effect of all these policies is to hold the total primary fuel demand slightly below the 1973/4 level right up to 2010 and beyond. The most important fuel-saving policies are those associated with transport and housing. The transport sector is analysed in detail in the Appendix. Here all that is necessary is to explain the basis of the calculations. Following this we shall look in some detail at the housing and food sectors before looking at the overall fuel demand.

The transport policy is based on a long-term goal of reducing the total passenger mileage. This runs contrary to the established expectations of increased personal mobility and can only be accomplished over a fairly long time-scale. The essential feature of the policy is that it starts to remove the need for personal mobility by changing planning policies. At the moment towns are planned in such a way that shops and workplaces are separated from residential areas. There is no reason in principle why planners should not reverse this policy and try to design towns so that houses, shops and workplaces are intermingled. However, even if *all* new developments were planned in this way there would not be any significant decrease in passenger travel for twenty-five to thirty years. Hence in the transport calculations total passenger mileage is reduced only after the year 2000. In addition to this long-term decrease the scenario also presumes that successful policies can be devised and implemented to cause a shift towards increasing use of public transport and, more significantly, an increase in the number of people travelling in each car. These assumptions, plus the efficiency improvements used in the t-f scenario, are sufficient to reduce the fuel demand of passenger transport to half the present level by the year 2010 (see Appendix for details).

In the 'house and home' sector the principal fuel-conservation measure is in the use of solar energy to provide about two thirds of the space and water heating for houses. In addition multi-unit dwellings (for instance high-rise tenement blocks) are heated by using part of the 'reject heat' of power stations. These policies, like the insulation policy of the t-f scenario, require an initial fuel investment and increase in the consumption of certain materials. However, like the insulation policies these policies repay the fuel investment within a year or two. There is however a very significant difference between the two policies, and the calculations which follow. The difference is in the state of our knowledge of solar and district heating systems. In calculating the fuel saving associated with double glazing I was able to use well-established figures for the fuel cost of the installation and for the subsequent fuel savings. To calculate either of these numbers for a solar conversion one has to rely on design data. It must be realized that these figures are not so reliable and that it is possible that either the fuel costs or savings may be substantially in error. My calculations are based on a house converted to solar heating in the new town of Milton Keynes. This is a fairly normal three-bedroom terraced house and, if the designers are correct, it will obtain 70 per cent of all its space- and water-heating requirement from a 40m$^2$ solar panel on its roof. The solar panel itself comprises a black aluminium panel separated from a covering sheet of glass. The aluminium panel has water channels in it which carry the heat from the panel to a very large central hot-water tank. This hot-water tank provides heat for central heating and for a conventional hot water system with a small gas boiler provided as a back-up system.

*Table 20  Inputs for 40m² solar collector*

|  | $10^3$ kWht |
|---|---|
| aluminium 8·5 kg/m² (at 80 kWht/kg) | 27·2 |
| glass 9·5 kg/m² (at 9·2 kWht/kg) | 3·5 |
| copper storage tank, 500 gallon capacity 300 kg (at 20 kWht/kg) | 6·0 |
| electrical equipment, extra pumps, valves, etc. (£75 at 50 kWht/£) | 3·7 |
| energy to install (£100 at 4·5 kWht/£) | 0·4 |
|  | 40·8 |
| Less tiles for 40m² roof area (45 kWht/m²) | 1·8 |
| *Net fuel cost* | $39·0 \times 10^3$ kWht |

The fuel costs of installing the 40m² solar-heating system are shown in Table 20. The total fuel cost is $39 \times 10^3$ kWht. If we assume that this provides 70 per cent of the space-heating and water-heating requirement then, once installed, it will save $18·9 \times 10^3$ kWht each year. (This is based on a fuel consumption per house of 33,800/kWht/year, of which 80 per cent is for space and water heating.) Thus the conversion pays for itself in fuel terms in just over two years. Like the conversion to double glazing described in the t-f scenario the solar-conversion policy will be constrained by the rate at which the outputs of the glass, copper and aluminium industries can be increased. Table 21 shows the size of the extra demand for different rates of conversion. I have assumed that the rate of solar conversion follows a curve similar to that used in the double-glazing calculation (Figure 47, p. 169) except that it reaches a peak in 1990 and falls after 1995. This means that by the year 2010 about 18 million houses are converted to solar heating with a net fuel saving of $340 \times 10^9$ kWh/year. In this same period the fuel demand for space and water heating would have increased by about this amount in a b-a-u scenario, as shown in Figure 57. Since I am assuming that people will still want most of the increased comfort associated with the b-a-u scenario, the solar conversions simply hold domestic fuel demand constant.

## 196  Fuel's Paradise

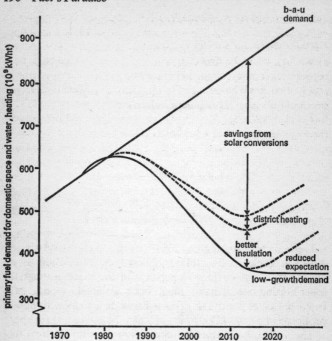

Figure 57   Fuel use for domestic space heating, showing how the demand in the b-a-u scenario is successively reduced by a series of policies.

Table 21  *Material requirements for solar conversions*

Each house needs 180 kg aluminium; 200 kg glass; 300 kg copper

|  | U.K. production (tonnes) | U.K. consumption (tonnes) | Extra demand converting 300,000 houses/yr (tonnes) | Extra demand converting 600,000 houses/yr (tonnes) |
|---|---|---|---|---|
| aluminium | 460,000 | 775,000 | 108,000 | 216,000 |
| glass (sheet) | 414,000 | 414,000 | 120,000 | 240,000 |
| copper | 180,000 | 700,000 | 180,000 | 360,000 |

There are, however, three other fuel conservation measures at work in this sector. The first is the use of district heating to supply 4 million dwellings with space heating. This is estimated to save about $30 \times 10^9$ kWh/annum when completed. In the same time period a substantial proportion of the housing stock will have been rebuilt with better thermal insulation. This will give an additional saving of $100 \times 10^9$ kWh/annum by 2010. It is not sensible from a fuel-conservation point of view to double-glaze and insulate any more houses, since the extra insulation on an existing house is mostly saving 'solar heat', not fossil fuel. (Extra insulation will only affect the fuel consumption when the solar system is not sufficient, but the insulation has to be produced using fuels.) Finally there is a saving due to 'decreased expectations'. If the overall policy of this scenario is successful, it would be unreasonable to let the fuel demand for home heating continue to rise indefinitely. Thus after 2010 the fuel demand curve is kept horizontal by assuming that everyone is happy with the level of comfort they would have had in a b-a-u scenario by the year 2010. These are all illustrated in Figure 57, which shows that the net fuel demand for space and water heating in this scenario is reduced from a 1975 value of $590 \times 10^9$ kWh to a value of $360 \times 10^9$ kWh from 2010 onwards.

The total fuel demand for the 'house and home' sector is not reduced so dramatically because the other parts of this sector, building houses, providing furniture, durables etc. are less affected by the scenario policy. I have assumed that house building is unaffected by the scenario but that the *growth* rate in the provision of household equipment is reduced from 2–3 per cent per annum to 1 per cent per annum. The total fuel demand for the sector is shown in Figure 58 along with those for the technical fix and business-as-usual scenarios. This shows quite clearly that the solar-conversion policy produces a significant fuel saving only after 1990 and would thus be considered a 'long-term' policy by present policy-makers.

The last sector which deserves detailed examination in this scenario is the food sector, since here there are also some fairly radical policies. The most radical policy advocated by those who support this scenario is that the U.K. should concentrate a far

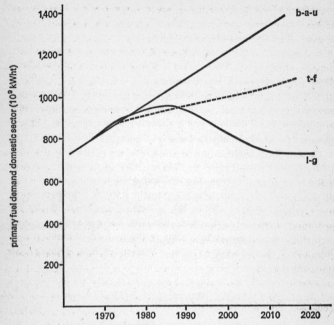

Figure 58  Total fuel demand for the domestic sector.

larger fraction of its agricultural resources to the production of fruit and vegetables and less to the production of meat and animal products. Although this is an important feature of a 'self-sufficiency' policy it is not of any significance from a fuel point of view. The production of potatoes in a field previously used as pasture for beef cattle may require more tractor fuel, but the marketing of potatoes is less energy-intensive than beef, and potatoes don't have to be fed on barley from the adjacent field. A shift of this type will provide a far larger food output for human consumption, but would require substantial changes in national dietary habits to be successful. It is therefore a long-term food-policy issue which can be ignored for our purposes.

Of greater significance is the policy of hydrolysing straw to provide fuels for farms and of reducing the use of fertilizers and packaging. The use of straw to provide an alcohol fuel is another technology which has been proved in principle but not widely applied. It has been shown that if this technology could be successfully developed it could provide *all* the fuels needed on a farm. In this scenario I have assumed that it is successful enough to reduce the growth in farm fuel use to zero by 1980 and cause a 15 per cent saving in the following two decades. For fertilizers and packaging I have assumed that total consumption can be reduced by 20 per cent by the year 2010. This is not a drastic reduction but is a more stringent policy than that used in the t.f. scenario.

The details of the changes in growth rates for the four decades 1970–2010 are shown in Table 22 in a similar way to that discussed in the t-f scenario. As then, I have allowed significant growth in the machinery and material input items to allow for technical improvement, reduction in electricity use etc. The overall effect of these policies is illustrated in Figure 59, which shows that after an initial growth the fuel demand levels out at about the 1974/5 level. For comparison the b-a-u and t-f curves show a fairly rapid growth over the same period.

*Table 22  Growth rates in the food sector*

| | Per cent of total fuel (per cent) | Growth in ten-year period (per cent) | | | |
|---|---|---|---|---|---|
| | | (b-a-u) (case) | 1970/80 | 1980/90 | 1990/2000 | 2000/10 |
| Hired transport | 3·2 | (10) | 5 | 2 | 0 | 0 |
| Farm fuel | 8·4 | (20) | 5 | 0 | −10 | −5 |
| Food, incl. fuel | 16·2 | (10) | 5 | 0 | 0 | 0 |
| Fertilizers, etc. | 11·3 | (15) | 0 | −5 | −10 | −5 |
| Packaging | 10·4 | (20) | 0 | −5 | −10 | −5 |
| Other in puts (machines, etc.) | 13·6 | (15) | 10 | 10 | 7·5 | 5 |
| Food-shop fuel | 15·9 | (35) | 10 | 5 | 0 | 0 |
| Food-shop machinery | 2·7 | (2) | 10 | 10 | 7·5 | 5 |
| Café, pub, etc. fuel | 13·0 | (35) | 10 | 5 | 0 | 0 |
| Café, pub, etc. machinery | 5·2 | (20) | 10 | 10 | 7·5 | 5 |
| Weighted averages (per cent) | | (21·1) | 6·43 | 2·57 | −1·39 | −0·43 |

200   *Fuel's Paradise*

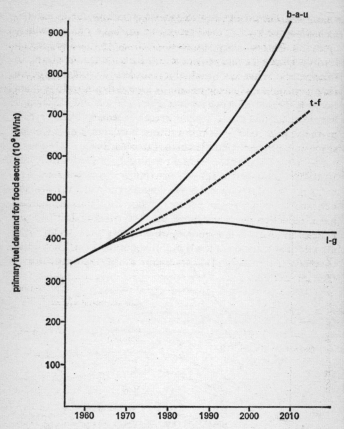

Figure 59   Total fuel demand for the food sector.

The three sectors which we have analysed in some detail, domestic, food and passenger transport, account for more than two thirds of total fuel demand in 1975. The other three sectors, public services, clothing and other goods and services, are only marginally affected by the low-growth policy. For example it is assumed that there is continued but reduced growth in the public-service sector (the reduction is because we are no longer building more motorways). The clothing sector is allowed to grow at 1–1·5 per cent p.a. instead of the 2·0 per cent p.a. in the b-a-u scenario. This reduction in fuel demand could be accomplished by reducing the quantity of synthetic fibres used in the clothing industry. The 'other goods' sector also grows more slowly than in the b-a-u case, at 1–2 per cent per annum instead of 2–3 per cent, owing

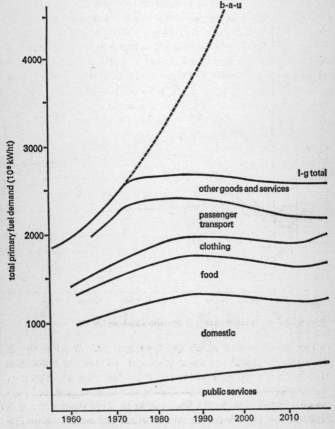

Figure 60  Total primary fuel demand for the l-g scenario.

202  *Fuel's Paradise*

to the reduction in use of electricity and some reduction in demand.

Although these three sectors continue to increase their fuel demand, the total primary fuel demand is held constant by the reduction made in the transport, food and housing sectors The total primary fuel demand is shown in Figure 60. It is noticeable that the three growth sectors (public services, clothing and 'other goods') increase their share of total fuel demand from 30 per cent in 1975 to 50 per cent by 2010. It is likely that were this scenario used as the basis of an energy policy then in time these growth sectors would also be curtailed so that fuel demand could be held constant fairly indefinitely.

As you might expect the significant reduction in total fuel de-

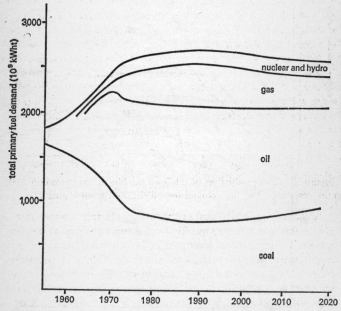

Figure 61  A possible supply option for the l-g scenario which requires only a small input from nuclear power.

mand realized in this scenario just about removes all the problems of matching fuel sources to demand. We saw that the moderate conservation policies in the technical-fix scenario opened up a whole series of options to policy-makers. In this scenario the supply options are more numerous. Since this is not a problem area in the scenario there is no point in exploring these options exhaustively. Figures 61 and 62 show how a non-nuclear option could make use of North Sea oil and indigenous coal supplies well beyond the end of the time-scale we are considering. It is interesting to notice that this is the only scenario in which North Sea oil is made to last us thirty years, yet it is also precisely the scenario rejected by those who expect North Sea oil to last us this long!

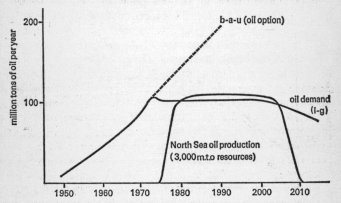

Figure 62   The production of oil from the North Sea (based on 3,000 m.t.o. resources) and the demand for oil deduced from Figure 61.

We have been able to analyse the fuel supply and demand situation in this scenario very quickly because we were able to use the principles and methods developed in the previous chapters. I am sure that you can also see how the political problems discussed in some detail in the technical-fix scenario will also apply here – only far more so. Because these problems are not illuminated or emphasized by the methods of energy analysis it is tempting to think that they are in some way less real and hence more easily overcome. To avoid this possible misinterpretation, I now want to spend a little time spelling out some of the non-fuel implications of this scenario so as to emphasize both the reality and the magnitude of the problems posed.

The first point which needs emphasizing in this scenario is that, like the b-a-u scenario, it depends on the successful implementation of technologies which have not yet been proved. In Chapter 10 the problem was with breeder reactors, here it is with solar roofs, straw hydrolysis units and public transport systems. A policy-maker who relies on these latter technologies is taking a risk of the same magnitude as one who depends upon breeder reactors, and the risk is not negligible. The significant difference between the scenarios lies, of course, in the consequences of failure. Here a failure to develop one or more of the technologies would mean an increase in fuel demand which could still be met from the available services. Furthermore the scale of the developments is very different. A prototype breeder reactor may cost £2,000 million and need back-up research costing as much again. The experimental solar house in Milton Keynes added a few thousand pounds to the cost of the house involved. (Another experiment in the same town, a 'dial-a-bus' public transport system, has cost tens of thousands of pounds to set up and try out.) These lower financial costs mean that the demand for all types of resources is very much less and so many different experiments can be attempted simultaneously. In contrast, nuclear developments have to concentrate on fewer and fewer systems as the financial and manpower costs increase.

There are, however, significant resource costs associated with the low technologies required for this scenario which may pose different types of problems. Perhaps the most striking are the material requirements for the solar conversion scheme. For instance to convert all 18 million houses to solar heating requires 3·25 million tons of aluminium – which is about one third of the present total *world* production of aluminium. Even allowing the conversion to be spread over a long period produces problems, as shown in Table 21. At the peak of the conversion programme, corresponding to 700,000 houses/year, the material demands would use up half the U.K.-produced aluminium, almost three quarters of the U.K. sheet-glass production and more than twice the U.K. copper production. The conversion policy will, of course, be managed so that the respective industries are expanded in concert with the rising demand – but this still means that the U.K. has to obtain larger quantities of raw materials from abroad.

If some of the events described at the beginning of the chapter were to happen, this would either cause a severe balance-of-payments problem or the conversions would not take place. In short the reduction in domestic fuel demand has here provoked a fairly serious problem by increasing the demand for some materials.

The other major policy which reduces fuel consumption is in the passenger-transport sector, and this raises another kind of problem. The drastic fuel savings are realized only by dramatically reducing the number of miles travelled by car and consequently the total number of cars. For this policy option to be successful the motor car must cease to be the symbol of personal status and cease to provide the basis of a major manufacturing sector of the economy. To a degree the very rapid increase in the price of petrol in 1973/4 has started the processes which are assumed to continue in this scenario. The number of new cars registered in 1972 was 1·9 million, in 1973 it was 1·75 million, in 1974 it was 1·3 million and it is estimated that it will be 1·1 million in 1975. In addition to this dramatic reduction in new car registrations there was a significant trend away from large U.K. cars to smaller foreign (particularly Japanese) cars. Hence the British Leyland bankruptcy in early 1975. At the time of writing, car sales have marginally increased, although still well below their seasonal norm, and car workers are facing the prospects of redundancies and short-time working. As explained in the Appendix, this scenario allows the car industry to continue to decline until by 1985 the output is reduced to 0·5 million cars, a quarter of the 1972 peak. If this were to come about, there would be a massive unemployment problem.

Firstly there would be the direct unemployment in the car industry itself. Then there would be unemployment in those materials industries which sell a significant fraction of their output to car manufacturers. As shown in Table 23 this corresponds to a total of at least 400,000 employees. If output is to be cut to a quarter of peak production, about 300,000 of these will be unemployed. However, this is not the end of the dismal story.

*Table 23  Employment in the car trade*

|  | thousands of men |
|---|---|
| Car manufacture | 200 |
| 8 per cent steel industry | 35 |
| 50 per cent asbestos industry | 15 |
| 50 per cent rubber industry | 70 |
| 25 per cent electrical machinery | 50 |
| other accessories | 50 |
| TOTAL* | 420 |

* Note that the total excludes the glass, copper and aluminium trades which are busy making solar panels etc.

In discussing the unemployment caused by a reduction in the electricity industry in Chapter 11 I pointed out that the capital industries (the industries which produce industrial plant and machinery) are sensitive to the *growth* in the economy, not just to total output. Thus a reduction in the *growth rate* from 2 per cent to 1 per cent halves the orders for capital equipment. In this scenario the capital industries are decimated by the cut-back on electricity generation, the decline of the motor trade *and* the general reduction in growth in all the other sectors. This means that a large proportion of those employed in the heavy materials and capital industries will be unemployed. This, in its turn, will reduce the demand for goods and services still further because people have smaller incomes when unemployed. The further reduction in demand will reduce the growth still further and cause more unemployment etc. This cycle of mutually reinforcing events is a positive feedback situation which can be very difficult to control. It was just such a sequence of events which caused the Great Depression in the 1920–30s, with well-known consequences.

The way out of the depression of the 1930s was, as first suggested by Keynes, for the government to increase employment by some social policy such as building roads or hospitals. In fact the only increase of employment large enough to break out of the 1930s depression was that brought about by the advent of the Second World War. Since the war the lessons of this economic disaster have not been forgotten and most Western governments have managed their economies so as to avoid a recurrence. What I am pointing out is that the low-growth scenario described here is a perfect recipe for reproducing the 'Great Depression' of the 1930s. It is therefore not surprising to find that most economists and government policy-makers totally dismiss this as a policy option.

The advocates of this scenario often fail to understand these reasons for dismissing the low-growth option. Those that do understand realize that the crucial link in the sequence of events which turns a reduction in growth to an economic disaster is unemployment, and this is the issue on which they concentrate. It has been suggested that the above discussion has overstated the unemployment problem and that it 'really isn't any worse than the unemployment caused by the run-down of the coal industry'. There are two simple answers to this comparison. In the first place the run-down of the coal industry was not done without causing severe social and economic problems. There is no doubt that the decline of the coal industry produced the 'depressed regions' in the U.K. with much larger unemployment problems than reflected in the national averages. Secondly, at the time when the coal industry was being run down the rest of the economy was 'booming'. There were many growth industries able to take up the manpower released from the coal industry. In contrast, in this scenario *all* the industries are reducing output or reducing growth; there is no expanding sector of the economy.

A more careful appraisal of the unemployment problem by advocates of this scenario rely on the trade-off between fuel use and manpower productivity to 'solve' the problem. In Chapter 3 I pointed out that the use of fuel has greatly enhanced our productivity per man. Well then, if we have too many men (that is to say, unemployment) and want to further reduce our fuel consumption, all we have to do is reverse the recent trends in technology and make our production systems more labour intensive and less manpower intensive. There are two problems here. The first is that a strong driving force towards lower labour intensity has been that individuals do not like to do monotonous hard physical work. Would you want to carry boxes across a warehouse all day, especially when you know that you could achieve ten times as much sitting at the controls of a fork-lift truck! The second problem is far more serious and poses a paradox within the context of this scenario. This is the relationship of labour intensity to unit costs and the implications this has for our ability to take part in world trade. There is no doubt that even in this scenario the U.K. needs to buy raw materials and food from abroad. It must therefore be able to sell some of its products abroad. However, if all our products are more labour-intensive than those of our competitors, they will cost a lot more and we shall be unable to sell them abroad. This could be avoided if all our competitors also decided to follow 'low-growth' policies *or* if the relative standard of living in the U.K. were to fall as fast as our labour intensity increased. The first of these possibilities is so improbable that it could not provide the basis of a policy. The second possibility would mean that if our labour intensity were doubled then our *real* standard of living would have to be *halved*. This is totally unacceptable to the present population of the U.K. It is also unlikely that *any* government would *ever* adopt a policy with such implications. Thus the paradox is complete: to try to make the U.K. more self-sufficient by adopting a low-growth fuel policy we first have to find a way of obtaining the necessary food and raw materials from abroad!

What this discussion has shown is that the social and economic problems associated with a low-growth scenario are not soluble by any of the techniques available to us at the moment. Leaving outside the very serious problems of persuading people to give up travelling by car and to reduce their material expectations, there are still very serious structural problems in the management of our economy which we are not equipped to solve. The problem is not simply one of unemployment; the problem hinges on a positive feedback situation in which a reduction in growth produces some unemployment which produces a further reduction in growth. This cycle of events, coupled with an overall policy of reducing fuel demand, leaves us with the choice of *either* lowering our standard of living (by factors of two, three or more) *or* doing without imports of food and raw materials.

# 13 The Energy Shop

We have now come full circle. At the beginning of this book we saw what an 'energy-conscious' society which had followed a 'low-growth' policy could be like. The last chapter explained most of the reasons why this is not a viable option for the U.K. – yet. In this chapter I want to bring together a number of themes and ideas which have occurred in various parts of the book so as to resolve this apparent contradiction.

My own interest in energy research began with the realization that climatic factors imposed a limit on the rate of use of fuels. This limit has not been of much concern to us in our investigations of future scenarios because it lies just outside the time-scale we examined. However, the existence of this limit leads to a number of important conclusions about the long-term direction of energy policies. We have seen how the production of all goods and services involves the consumption of fuel and that in the future some of these may require more fuel and some less. In the long term there is a limit to how much technology can be improved to reduce fuel consumption. Since, also in the long term, all resources will become leaner (lower grade) the overall trend will be for more fuel to be needed for the production of the same output of goods and services. This puts mankind in a type of 'energy shop' where his annual expenditure is fixed (by the climatic limit) but the price (fuel cost) of all the commodities in the shop is gradually increasing. This is not a comfortable view of the future, since the choice implied, of either exceeding the climatic limit or lowering standards of living, does not have any desirable outcome.

There is a way out of the energy shop problem. It hinges on the fact that mankind has, in addition to a stock of fuels, an income of energy – principally solar energy. If solar energy is used to provide some or all of the goods and services, the climatic limit does not impose such a stringent limit on our annual expenditure. This is because solar energy, and other income sources such as wind, tides etc., already contribute to the energy flows in the atmosphere. Thus if man makes use of these energy flows, and subsequently puts the energy back into the atmosphere, he has not altered the total energy flow. There may be second-order effects according to where and when the energy is put back into the atmosphere, but these are a lot less significant than changes which could result from a net addition of energy to the atmospheric system. So if we start to make use of these income energy sources we avoid the problem of a limited expenditure in the energy shop.

We have already seen an example of this type of change in the use of solar energy in the low-growth scenario. The important feature of the analysis of domestic fuel demand in that scenario was that although the *fuel* demand was significantly reduced the amount of heat *energy* used in all the houses was exactly the same as in the other two scenarios.[1] In other words if we were to follow a 'low-growth' policy we would all be just as warm and comfortable as following other policies, but our demand on fuel resources and our impact on climate would be much less. In the *long term* this seems to me to be the only sensible direction for energy policy.

Earlier I have emphasized the importance of the technological inertia, or time-lag, in our industrial system. The time-lag is about twenty to thirty years and arises because of the scale and complexity of our technologies. For example, the U.K. nuclear programme was begun in the early 1950s with the construction of Calder Hall, yet by 1974 nuclear power was only supplying 3 per cent of the primary fuel input to the U.K. We have also seen this in action in our scenario explorations. For example, the 'crash programme' employed to insulate houses in the technical-fix scenario put a severe strain on the glass industry and was able to insulate only

---

1. It is because I attach importance to the climatic limit and being able to avoid it by using income energy sources that throughout the book I have been careful to distinguish between fuel and energy. This is also why I count the primary fuel equivalent of nuclear and hydro power in terms of the 'heat released'.

90 per cent of U.K. houses by the year 2010. This time-lag means that if we want a significant fraction of our *energy* supplies to come from equilibrium sources by about 2010–2020 then we have to take policy steps *now*. However, in discussing both the technical-fix and low-growth scenarios I spent some time pointing to political and economic factors which could be regarded as a sort of 'social time-lag'. The effect of these factors is that if a policy-maker wants, say, to reduce the growth in electricity demand, he has to prepare the ground in political and economic ways. This may require starting campaigns to reduce the sales of electric heating devices (a massive tax?) and finding other ways of employing the resources of the electrical engineering industry (making heat pumps?). To avoid serious political confrontations or an economic recession these policies should be pursued slowly, say over a five-to-ten-year period. This social time-lag has to be added on to the technological time-lag, giving a total between twenty-five and forty years. This is now a long-term consideration.

These arguments lead me to the conclusion that we should move towards a low-growth policy, which utilizes income energy sources, but that it is *only* a long-term option. The social, economic, technical and political conditions prevailing in the U.K. now mean that an immediate low-growth policy is not possible. This does not, however, mean that steps should not be taken now so as to make this a viable policy option at some time in the future.

This conclusion emphasizes one of the serious drawbacks implicit in the 'scenario' approach to formulating policy. Throughout the evaluation of any scenario I kept the same set of assumptions over a long period of time; in other words, I stuck to one policy. This is a sensible procedure if you want to see the implications of one policy carried to its end, but it tells you nothing of all the options which arise if a policy-maker decides to change his policy at some point in the future. There are other problems with the scenario approach which are important if you want to draw any conclusions from these analyses. Firstly, in a finite length of time you can examine only a few scenarios in any detail. I chose to examine three options and chose them to span a significant range of future options. But of course these are not the only possibilities. There are energy policies which lie between those characterized as

## 212 Fuel's Paradise

b-a-u and t-f and policies between t-f and l-g. Secondly, I have described each of these policies in terms of a single fuel-demand curve, as shown in Figure 63. This is sensible from an analysis point of view. However, there are many uncertainties in the future, and the fact that I have chosen a particular number as the outcome of a policy hides the fact that other estimates would arrive at differ-

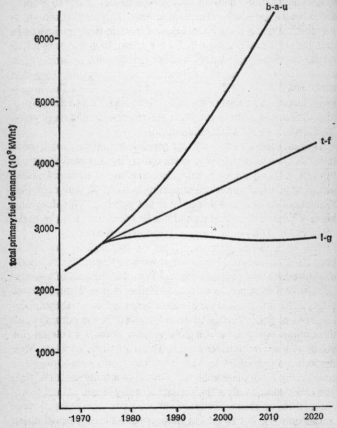

Figure 63   The total primary fuel demand for the three scenarios.

*The Energy Shop* 213

ent results. Furthermore, as we go further into the future the degree of uncertainty becomes larger and larger. Thus a more realistic representation of the three scenarios would be to draw three overlapping shaded regions, as in Figure 64, not three neat curves. This is not simply an academic point. It is important because it shows that under some conditions there might be no

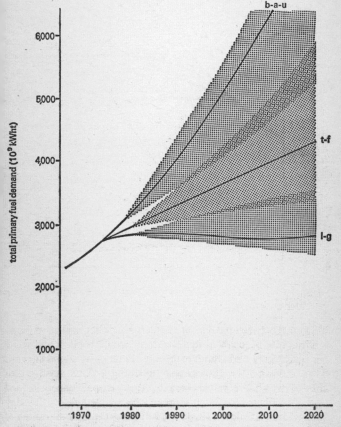

Figure 64 The shaded areas emphasize the increasing uncertainty with increasing projection time.

214 Fuel's Paradise

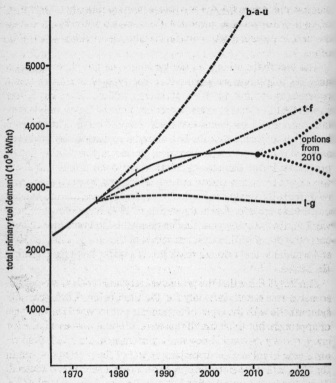

Figure 65  A plausible trajectory which starts close to the b-a-u curve but later changes to an l-g curve.

difference between the *outcomes* of a b-a-u policy and t-f policy.

What all this means is that the future is not as black and white (or black, grey and white) as a scenario analysis would suggest. There are not a few choices open to us but a whole spectrum of choices which are represented by the trumpet-shaped shaded region of Figure 64. What is more, there is no reason why any society should follow a single smooth curve through this range of options. The U.K. could choose to follow a curve close to b-a-u for ten years, then opt for a lower-growth curve for ten years and so on, as shown in Figure 65. At each future point in time there is a range of options open and, provided changes are introduced slowly enough, there is no reason why policies should not be changed fairly often.

The fact that nothing is quite as simple as the three scenario analyses might suggest means that this type of exercise doesn't provide you with a ready-made energy policy. However, it has shown that the 'energy crisis' experienced in 1973 was only a curtain-raiser to the problems that could descend on us. The analyses have also suggested that the two extreme scenarios, the business-as-usual and low-growth scenarios, present a formidable list of problems. In the business-as-usual scenario we found that there was only one energy policy option, which was to push every fuel resource to its maximum production as fast as possible and to keep our fingers crossed. Again the events of 1973 have shown policy-workers that keeping your fingers crossed is no guarantee of success. The list of problems summarized at the end of Chapter 10 is so formidable that I do not think this is a viable basis for planning the future.

Similarly I think that the problems associated with a low-growth scenario rule it out, but only for the time being. I am generally sympathetic with the types of arguments put forward for this type of approach, but think that if this were adopted now as a basis for energy policy we would bring more problems on our heads than we are trying to solve. One disturbing aspect of the present debates in the U.K. is that these are the only two scenarios seriously discussed. There has been a polarization between the 'hard-line-fixers' of the CEGB and the 'eco-nuts' from Friends of the Earth. Both choose to ignore the problems associated with their chosen policies. This type of polarization is unfortunate because it means that anyone advocating a middle road will be severely criticized by both of the most vociferous parties. In effect the polarization puts us in the dangerous position of being forced to choose between two potentially disastrous policy options.

This suggests that we should be looking for an energy policy of the 'technical-fix' type. However, the discussions in Chapter 11 pointed out that this type of policy was no panacea and involved its own set of technical and political problems. There are three reasons why I think that this should form the basis of present U.K. policy. Firstly, the idea of advocating any policy which did not attempt to achieve maximum technical efficiency goes against all the value judgements ingrained in me by my scientific training. If you have difficulties finding adequate fuel supplies, it seems absolutely criminal to allow or even encourage the waste of fuels by heating homes electrically or driving 3-litre cars around towns when 1-litre cars do the same job. Secondly, of all the policy options open to us this is the policy which gives future policy-makers the most options. Essentially the technical-fix policy is using better technology to buy time. Later, say by 1990, many of the factors which are now uncertain may be resolved and it may be a lot clearer which direction energy policy should take. Finally, it seems obvious to me that in the long-term (50–100 years) we do have to reduce our rate of growth in fuel consumption, and this policy is setting off in that direction as fast as is presently feasible.

There are some signs that the U.K. government is moving in the direction of a 'technical-fix' policy. A recent publication by the National Economic Development Office (NEDO, 1974) sets out a number of policy options which could reduce fuel consumption. These are more moderate options than those advocated in Chapter 11 in the sense that they are less likely to arouse political antagonism. However, the aim is to reduce fuel consumption by about 10 per cent by 1985, which is a smaller reduction than that resulting from the policies discussed in Chapter 11. There are also many official bodies still advocating a b-a-u option. The OECD publication *Energy Prospects to 1985* sees the problem as one of substituting some other fuel for oil which is now too expensive, but it does not advocate a deliberate reduction in overall fuel demand. A meeting at the Royal Society in May 1974 ('Energy in the 1980s') brought together a powerful group of individuals each of whom was advocating an alternative source of fuel to replace oil.

The danger with a technical-fix policy is that it is a 'compromise solution', a 'middle road', and, as such, is unlikely to arouse the level of emotional commitment which the other scenarios are able to evoke from their respective advocates. This means that it may become a half-hearted and rather feeble energy policy. This is especially likely in the U.K., since North Sea oil will produce an 'energy bonanza' in the 1980s. But as the analyses have shown North Sea oil is *only* a bonanza, it does not provide a basis for a long-term energy policy. Thus it will require considerable determination and a fixed stare on the problems likely to arise in the 1990s to make such a technical-fix policy work.

Throughout all the discussions of the future I have emphasized the degree to which our future is a matter of our choice. This is an oversimplification. No matter how much choosing you do you cannot change the laws of thermodynamics, nor can you choose to make the OPEC countries reduce the price of oil, nor can you choose the U.K. to have a warmer climate. It is easy enough to incorporate factors such as the English weather into your policy-making (you simply assume the worst). However, as pointed out earlier, there is no way of accommodating factors such as the state of world trade, the possibilities of world wars or lesser events, such as shortages of uranium, into policies. These *are* all crucially important factors and, as spelt out in all the scenarios, some of them will make some policy options impossible. The sheer difficulty of keeping track of the options made possible and impossible by world events has had the effect of reducing the time horizons of government policy-makers. Put crudely they are so busy working out what is possible over the next five years that they have not got time to look over a thirty-year period.

To a large degree this has led to a philosophy of determining policy decisions by doing what you *can* do rather than what you *want* to do. This means that policy options are chosen more on the grounds of whether the problems posed by the option can be solved than on the grounds of whether the outcome of the policy is a desirable one. This is a sophisticated way of saying that modern government is more concerned with 'means' than with 'ends'. The implication is that a government choosing between a b-a-u option and a t-f option will always choose the b-a-u option because it involves only technical problems whereas the t-f option involves political problems.

Over the last fifty years our society has had a remarkably good record of solving technical problems – and a remarkably bad record of solving political problems (for example Northern Ireland, distribution of income, etc.). Of course the fact that we have solved technical problems in the past does not guarantee that we shall be able to solve the technical problems of the future. But if you are guided mostly by what you *can* do, you will choose an option which poses problems you *think* you can solve rather than an option which poses problems you think you cannot solve.

In essence my argument for changing this approach rests on the belief that the climatic limit represents a technical problem which is in principle not 'soluble'. This is not because I think it is a very difficult problem; it is because the laws of thermodynamics, which have been well verified over the last hundred years, preclude any possible solution to the problem.[3] Faced with a technically insoluble problem, the approach to energy policy must change. We now have a goal, avoiding the climatic limit, which is more important than any of the 'means' of implementing policy. This means that our efforts now have to be directed towards solving the problems which will enable us to reach the goal of our choice.

This will not be easy. I do not have any solutions to the economic and political problems which are associated with the reduction in the growth of fuel demand, nor do I know where solutions can be found. I fear that the problems are likely to be made worse by world events and that problems will arise which cannot be foreseen now. But being difficult to solve is much easier than being impossible to solve.

This goal, avoiding any significant change in our climate, requires us to change the direction of technological development.

3. Some people have argued that it is short-sighted to use the laws of thermodynamics. They argue that before Einstein I would have been wrong to draw conclusions from Newton's laws. As soon as thermodynamics has its 'Einstein', we shall find a way round the problem. This type of argument shows a fundamental misunderstanding of the ways in which knowledge in physics changes with time. The introduction of Einstein's relativity did not change the way that planets move, nor the way apples fall to the ground, nor the way that cars move along roads. In all these situations Einstein's theories predict exactly the same result as Newton's theories – as they must, since the same events take place. Einstein's theories only changed our views of motion on a very small scale (atomic) and on a very large scale (astronomic). In exactly the same way any revolution in the theories of thermodynamics may change our understanding of some circumstances, but they will still predict that putting more heat into a body makes it hotter, and that the total energy into a system will equal the total energy out.

It does not require us to 'go backwards', but it does require a new evaluation of what does and does not constitute 'progress'. Until comparatively recently the growth of human populations has pushed technology in the direction of improving productivity to compensate for declining resources (Wilkinson, 1973). This has been especially noticeable in agriculture, where every increase in population has required an equivalent increase in output per acre. The population of the U.K. is now fairly constant and without any draconian curtailment of personal liberty could even be slightly reduced. The trouble is we have come to define 'progress' as meaning more or bigger or faster and so continue consuming resources in ever larger quantities. The U.K. could, over a sensible period of time, redefine the conventional definition of progress to be closer to 'sustainable' than to 'more'. This would at least reduce our growth in resource consumption and go some way towards allowing the non-industrial countries to achieve something resembling a comfortable existence. But this can only come about if we can control the direction of our development, in particular the direction of our energy policy.

Until now U.K. energy policy has been blown about by the winds of world events. It has been like a very crude sailing ship which can only sail with the wind. This has not been too serious since until now the most important criteria have been to keep afloat and to keep moving. Our technologies have enabled us to increase the sail area and move faster and faster with the wind. But now there is a storm up ahead and the wind is blowing us straight into it. To sail into the storm with our present understanding of boats and winds spells certain disaster. So we have got to get really smart and invent a rudder for the energy policy ship. Initially it will be a very crude rudder and it will only roughly determine our direction of movement. When the wind changes we will get blown off course, first one way and then the other. But with a rudder we can at least set a course and aim to avoid the storm. Eventually we may have to get clever enough to be able to sail into the wind in order to keep ourselves headed in the right direction. Maybe when we can do that we shall be able to avoid not only the storm centre but other troubled waters. But first someone has to invent the rudder.

# Appendix: Fuel Use in Transport

In 1968 the use of fuels in providing passenger transport accounted for more than 20 per cent of all fuel use. Historically this sector is one of the fastest-growing sectors, especially in its use of fuel. It is therefore important to try to obtain as good an estimate as possible of future fuel consumption in providing passenger transport. There are many factors which make any kind of extrapolation into the future difficult. One of the most important reasons for including this appendix is to show all the assumptions which have to be made and how they have been dealt with. Presenting the calculations in this detail also allows you to insert your own assumptions into the calculations and see how much this affects the final result.

The largest single fuel consumer in the transport sector is the motor car. The fuel used in motor cars accounts for almost 10 per cent of total U.K. fuel use – almost half the total fuel used by the transport sector. To a large degree the growth or decline of the private motor car determines what happens to the rest of the passenger transport sector. By comparison with motor cars, buses and trains have a trivial fuel consumption. The analysis which follows is broken up into a series of stages as follows:

(i) The total passenger-mileage needed for each year is estimated. This is a measure of personal mobility.
(ii) The split (called the modal split) of this total passenger-mileage between cars and other forms of transport is then determined.
(iii) Next, the average number of passengers per car is estimated. Coupled with the total passenger-miles travelled by cars, this gives us the number of car-miles travelled.

(iv) The energy consumption per car-mile is calculated taking into account any changes in engine capacity and engine efficiency.
(v) The passenger-miles not done by car are then divided between bus, rail and air and the fuel implications calculated.
(vi) The fuel needed to produce cars/buses/railways etc. is assumed to be proportional to the passenger-miles travelled in each mode. This enables the indirect fuel consumption to be calculated.

In this analysis I will examine each parameter for each of the three scenarios described in the book. This has the advantage of allowing a fairly direct comparison to be made of the personal mobility implications of each of these fuel-policy data sources options.

The first task is to estimate the total number of passenger-miles desired in the future. This is a type of guessing game with a twist to it – the twist being that official guesses of personal mobility are to a large measure self-verifying. In planning new towns, road networks and public-transport systems planners guess how much people want to travel and design the system accordingly. Since the plans include factors such as separating homes from shops and factories, they are to a large degree self-justifying. Thus the starting point for our analysis is the official projections of passenger travel. An average of several such projections is shown in the top curve of Figure 66. It is reasonable in the sense that it shows some decrease in the rate of growth due to saturation. There is a limit on how many hours a day an individual can spend travelling!

Since both the b-a-u and t-f scenarios are based on continued satisfaction of consumer demand this extrapolation is taken as the passenger mileage in both these scenarios. In contrast the low-growth (l-g) scenario is based on deliberate changes in life-style, which include things such as moving people closer to their place of work and putting shops within walking distance of homes. Thus the passenger-mileage in the l-g scenario is assumed to stay fairly constant for twenty–thirty years and then, as the changes start to take effect, to decline to levels of personal mobility typical of the 1950s.

The next parameter to be estimated is the proportion of this total passenger transport which is done by motor cars. The percentage of car transport has been rising fairly steadily over the past twenty

## 222 Fuel's Paradise

years, and in the b-a-u scenario this trend is continued until by about 2010 car transport accounts for 97 per cent of passenger transport. In the t-f scenario the trend is substantially reduced and reaches a peak of 83 per cent of all journeys by car in 2005. In the l-g scenario the trend is reversed so that from a peak of 80 per cent car journeys in 1985 the percentage of journeys done by car has fallen to 50 per cent by 2015. Combining these trends in the percentage of journeys by car with the total passenger miles gives the passenger mileages by car shown in Figure 67.

In both the b-a-u and t-f scenarios it is assumed that the average number of passengers per car remains at today's value of 1·85. This average occupancy of cars combines two components, namely journeys to work, with an average occupancy of about 1·5, and holiday and weekend journeys, with an average occupancy of about 2·5 (H.M.S.O., 1969). Since 1963 there has been a larger increase in work journeys than in leisure journeys and the overall car occupancy has decreased from 2·03 in 1963 to 1·85 in 1973 (Highway Statistics). In the low-growth scenario it is assumed that work journeys are made less necessary and where required done by public

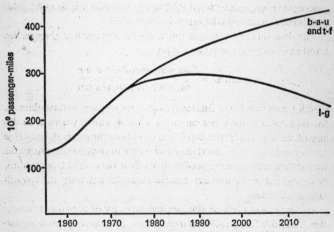

Figure 66  The historical trend and projections of total passenger mileage for the three scenarios.

*Appendix* 223

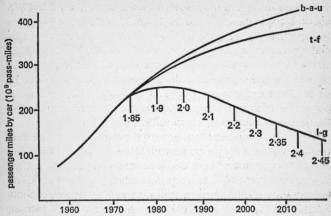

Figure 67 The projections of passenger miles travelled by car. The numbers alongside the l-g curve indicate the projected car occupancy.

transport. Thus, as the changes in life-style take effect, it is assumed that the average car occupancy in the l-g scenario rises towards 2·5 passengers/car. This trend of increasing occupancy is shown by the numbers alongside the l-g curve in Figure 67.

The data on car passenger-miles and car occupancy gives us the number of car-miles travelled, since

$$\text{car miles} = \frac{\text{passenger-miles by car}}{\text{number passengers/car}}$$

Finally, to estimate the fuel used in the cars we have to have data on the fuel consumption per car-mile. This of course varies with the size of the car, the traffic conditions and how the car is driven. It is not sensible to try to build into any scenario detailed assumptions on matters such as how people drive their cars. What we can do, however, is try to guess the changes in engine capacity and overall engine efficiency.

In the period 1960–68 the average capacity of registered cars in the U.K. fell from 1,400 cc to 1,300 cc, principally owing to the introduction of the 'mini' and other small cars. After 1968 the trend was reversed, until by 1973 the average engine capacity was again 1,400 cc. This shows that the average engine capacity can change quite quickly, even when there is a large stock of cars in existence. I have assumed that the recent trend of increasing engine capacity is continued in the b-a-u scenario so that by about 2005 the average engine capacity is about 2,000 cc, as shown in Figure 68. In the low-growth and technical-fix scenarios the 1960–68 trend of decreasing engine capacity is assumed to occur, so that by about the year 2000 the average engine capacity is 1,100 cc. This data can

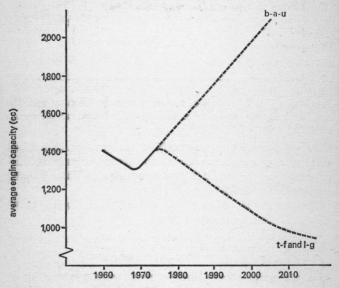

Figure 68 The historical and projected changes in average engine capacity.

now be combined with the data on fuel consumption and engine capacity shown in Figure 69. I assume that in the b-a-u scenario the fuel consumption follows the best-fit line shown in Figure 69. In the t-f and l-g scenarios I have assumed that there is an improvement in average engine efficiency, so that by the year 2000 the relationship between fuel consumption and engine capacity is the same as the dotted curve in Figure 69. Together these assumptions mean that whereas now the average fuel consumption is about 1·6 kWht/car-

*Appendix* 225

*Table 24 Breakdown of fuel consumption in the passenger transport sector*

|  | $10^9$ kWht |
|---|---|
| Car fuel | 212 |
| Car manufacture | 69·3 |
| Car parts (incl. garage equipment) | 35·0 |
| Motor service trades (fuel consumed in garages/insurance offices etc.) | 22·0 |
| Bus fuel | 14·4 |
| Railway fuel | 18·8 |
| Aircraft fuel | 50·6 |
| Ship fuel | 1·7 |
| Manufacture ships/buses/planes, etc. | 32·0 |
| Other transport terminals (airports, railway stations etc.) | 10·0 |
| TOTAL | $465·8 \times 10^9$ kWht |

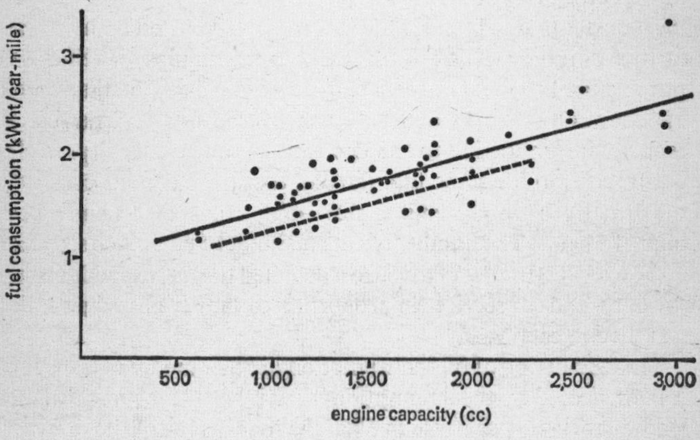

Figure 69 The variation in fuel consumption per car-mile as a function of engine capacity. The scatter arises because of different speeds, traffic conditions etc. (Source: Mortimer, 1975.)

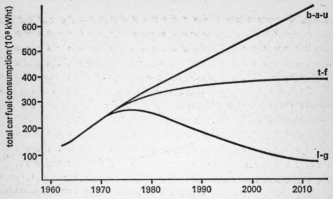

Figure 70  Car fuel consumption in each of the three scenarios based on the projections in Figures 67-69.

mile by the year 2000 it is 2·1 kWht/car-mile in the b-a-u scenario and 1·3 kWht/car-mile in the t-f and l-g scenarios. Combining these fuel-consumption values with the total car-miles produces the car-fuel consumption trends shown in Figure 70.

To obtain a total figure for passenger transport we have now to incorporate all the other items shown in Table 24 with their 1968 fuel consumption. To calculate the fuel consumption in the manufacture of motor vehicles I assume that the number of vehicles is proportional to the number of car-miles travelled. Furthermore I assume that in the t-f and l-g scenarios the fuel cost per car is reduced in proportion to the decrease in engine capacity. In the b-a-u scenario I assume that the fuel cost per car stays constant, the penalty of larger engine size being offset by technological improvements. The number of cars manufactured in each scenario is shown in Figure 71. I have also calculated the fuel expenditure on car parts and car services in proportion to the car passenger-miles travelled in each year.

The calculation of fuel consumption in other modes of transport is complicated by the assumptions which have to be made about load-factors. For example in 1968 the load-factor of the railways was only 16 per cent, indicating that the same rail system could carry eight times as many passengers without increasing its fuel

*Appendix* 227

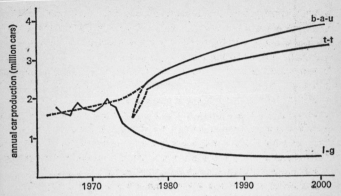

Figure 71 Annual car production projections based on the utilization of cars in each scenario.

consumption at all. In all the scenarios there are only small changes in the total passenger mileage travelled by other transport modes. Thus this is not a significant factor in changing the total fuel consumed. To make an estimate I assumed that passengers not travelling by car split their journeys 75 per cent to bus and 25 per cent to

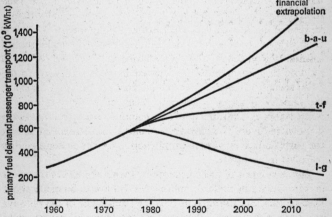

Figure 72 Total fuel demand for passenger transport in each scenario. Also shown is a projection based on estimated expenditures.

rail. The increase in rail traffic was accommodated by allowing the average load-factor to rise from 16 per cent to 24 per cent. The extra bus traffic needed 15 per cent more bus-miles and a slight increase in load-factor. In the b-a-u scenario I assumed that there was also a significant growth in air traffic which did increase the total fuel consumption.

The net result of all this number-crunching is the total transport fuel demand shown in Figure 72. As well as the three scenario projections I have also included the projection based on the simple financial expenditure procedure described in Chapter 10. The fact that this is higher than the b-a-u estimate arises, I think, because even in the b-a-u scenario the rate of growth of passenger transport is not as large as historical trends would suggest.

# Suggestions for Further Reading

The seven titles singled out below are not supposed to represent the 'best' from the fifty or so books on energy published in 1974. They have been chosen because they are a useful supplement to the themes developed in this volume. The first four give different views on U.K. energy policy and the last three describe similar exercises for the U.S.A.

*Energy and the Environment:* Report of a working party set up by the Committee for Environmental Conservation, Royal Society of Arts and Institute of Fuel, July 1974, Royal Society of Arts, London. Takes a much wider look at the environmental effects of past, present and future sources of fuel in the U.K.

*Energy in the 1980s:* Vol. 270 of the *Philosophical Transactions of the Royal Society, London,* May 1974, pp. 405–620. This collection of papers by recognized authorities provides a view of the 'official' view on energy resources and their use. The general approach provides an example of the 'business-as-usual' philosophy.

*Energy Conservation in the U.K.:* report by NEDO, H.M.S.O., 1974. This is a review of energy use in the U.K. with special emphasis on the potential for energy conservation. In many respects it is close to the 'technical-fix' scenario developed in this book.

*World Energy Strategies: Facts, Issues and Opinions:* by A. B. Lovins, Earth Resources Ltd (9 Poland St, London), 1973. Although set in the context of world supply and demand of fuels this contains a lot of data pertinent to the U.K. The case argued for is close to the 'low-growth' scenario in this book.

## 230  Fuel's Paradise

*Exploring Energy Choices:* preliminary report of the Energy Policy Project, Ford Foundation, Washington, D.C., 1974. An interim report of a very large study of energy options in the U.S.A. based on the 'scenario' approach. In many respects it is the precursor to this book, though based entirely on the U.S.A.

*U.S. Energy Prospects:* National Academy of Engineering, Washington, D.C., 1974. Presents a well-argued engineering analysis of the potential for increasing fuel supplies in the U.S.A.

*Energy: Demand, Conservation and Institutional Problems:* ed. M. S. Macrakis, M.I.T. Press, 1974. A volume of collected papers which discuss U.S.A. energy policy from economic, technical and energy analysis viewpoints. Very technical but full of information and ideas.

# References

BECKERMAN, W. (1974), *In Defence of Economic Growth*, Jonathan Cape.

BOUSTEAD, I. (1974), 'Resource Implications with Particular Reference to Energy Requirements for Glass and Plastic Milk Bottles', *Journal of the Society of Dairy Technology*, 27, (3), p. 159.

CHAPMAN, P. F. (1974a), 'The Energy Cost of Fuels', P. F. Chapman, G. Leach and M. Slesser, *Energy Policy*, 2 (3), September, p. 231.

(1974b), *Energy Analysis of the Census of Production 1968*, P. F. Chapman and N. D. Mortimer, Open University Research Report ERG 006, August 1974 (published September 1975).

(1974c), 'The Energy Required to Produce Copper and Aluminium from Primary Sources', *Metals and Materials*, February, p. 107.

(1974d), *The Energy Inputs and Outputs of Nuclear Power Stations*, P. F. Chapman and N. D. Mortimer, Open University Research Report ERG 005, September (see also *New Scientist*, 19 December 1974).

(1974e), 'Energy Conservation and Recycling of Copper and Aluminium', *Metals and Materials*, June, p. 311.

(1975), 'The Energy Costs of Materials', *Energy Policy*, 3 (2), March, p. 47.

DEPARTMENT OF ENERGY (1974, 1975), *Development of the Oil and Gas Resources of the United Kingdom*, H.M.S.O.

'ENERGY IN THE 1980s' (1974), Collection of papers following a discussion organized by Sir P. Kent, *Philosophical Transactions of the Royal Society of London*, 276 (1261), pp. 405–620.

E.P.P. (1974), *Exploring Energy Choices*, a preliminary report published by the Energy Policy Project, The Ford Foundation.

GALBRAITH, J. K. (1972), *The New Industrial State*, André Deutsch (Penguin Books, 1974).

GRIFFITHS, C. I. (1974), *The Meteorological Magazine*, March.

H.M.S.O. (annual), *Highway Statistics*, Department of the Environment.

(annual), *United Kingdom Energy Statistics* (previously *Digest of Energy Statistics*).

(1966), *A Century of Agricultural Statistics 1866–1966* (published 1968).

(1968), *Family Expenditure Survey* (also other years).

(1969), *Private Motoring in England and Wales*, P. G. Gray.

(1971), *Report on the Census of Production 1968*, 156 vols.

LAMB, H. H. (1970), 'Volcanic Dust in the Atmosphere', *Philosophical Transaction of the Royal Society of London*, A266 (1178), pp. 426–533.

LAWTON, J. H. (1973), 'The Energy Cost of Food Gathering', in *Resources and Population*, ed. B. Benjamin et al., Academic Press.

LEACH, G. (1974), 'The Energy Costs of Food Production', in *The Man–Food Equation*, ed. A. Bourne, Academic Press.

(1975), Private communication.

LEES (1970), 'Waste Heat in the Los Angeles Basin' (ref. S.C.E.P.).

LOVERING, T. S. (1969). 'Mineral Resources from the Land' in *Resources and Man*, W. H. Freeman.

MACHTA, L. (1971), 'The Role of the Oceans and Biosphere in the Carbon Dioxide Cycle', *Proceedings of the 20th Nobel Symposium*, Goteborg, Sweden, p. 121.

MITCHELL, J. M. (1970), 'A Preliminary Evaluation of Atmospheric Pollution as a Cause of the Global Temperature Fluctuation of the Last Century', in *Global Effects of Environmental Pollution*, ed. S. F. Singer, D. Reidel, Holland.

N.E.D.O. (1974), *Energy Conservation in the U.K.: Achievements, Aims and Options*, H.M.S.O.

ODUM, H. T. (1971), *Environment, Power and Society*, J. Wiley.

O.E.C.D. (1975), *Energy Prospects to 1985*.

MORTIMER, N. D. (1975), *Energy Costs of Road Transport*, Open University Research Report ERG 009 (in preparation).

PETERSON, J. T. (1969), *The Climate of Cities*, National Air Pollution Ad. Publication No. AP–59, U.S. Department of Health, N. Carolina, October.

PIMENTEL, D., et al. (1973), 'Food Production and the Energy Crisis', *Science*, 182, p. 443.

PRICE, J. (1974), *Dynamic Energy Analysis and Nuclear Power*, Friends of the Earth (9 Poland Street, London W1).

PUTNAM, P. C. (1953), *Energy in the Future*, U.S. Atomic Energy Commission.

ROBERTS, P. C. (1974), *A Method of Projecting Energy Demand in the U.K.*, P. C. Roberts and V. E. Outram, Department of the Environment.

SAHLINS, M. (1972), *Stone Age Economics*, Tavistock.

S.C.E.P. (1970), *Man's Impact on the Global Environment*, report of the Study of Critical Environmental Problems, M.I.T. Press, 1970.

SELLERS, W. D. (1969), 'A Global Climatic Model Based on the Energy Balance of the Earth-Atmosphere System', *Journal of Applied Meteorology*, 8, pp. 392–400.

SLESSER, M. (1973), 'Energy Subsidy as a Criterion in Food Policy Planning', *Journal of the Science of Food and Agriculture*, 24, p. 1193, 1975.

S.M.I.C. (1971), *Inadvertent Climate Modification*, report of the Study of Man's Impact on Climate, M.I.T. Press.

STEINHART, J. S. (1974), 'Energy Use in the U.S. Food System', J. S. and C. F. Steinhart, *Science*, 184, April, p. 307.

U.N. STATISTICAL YEARBOOK (annual), United Nations Organization.

VAUGHAN, R. D. (1974), *Uranium Conservation and the Role of the Gas-Cooled Fast Breeder Reactor*, paper presented to the British Nuclear Energy Society, January 1975.

WASH–1139 (1974), *Nuclear Power Growth, 1974–2000*, U.S. Atomic Energy Commission report, U.S. Printing Office.

WASHINGTON, W. M. (1971), 'On the Possible Use of Global Atmospheric Models for the Study of Air and Thermal Pollution', in *Man's Impact on the Climate*, ed. W. H. Matthews et al., M.I.T. Press, pp. 265–76.

(1972), 'Numerical Climatic-Change Experiments: The Effect of Man's Production of Thermal Energy', *Journal of Applied Meteorology*, 11 (5), pp. 768–72.

WEINBERG, A. M. (1974), 'Global Effects of Man's Production of Energy', *Science*, 18 October.

and Hammond, R. P. (1970), 'Limits to the Use of Energy', *American Scientist*, 58, pp. 412–18.

WILKINSON, R. C. (1973), *Poverty and Progress*, Methuen.

# Index

atmosphere, energy in, 77
  models of, 82

car, engine capacity projections, 224
  fuel consumption of, 223–7
  industry, 205–6
  production of, 227
carbon dioxide, 80
chemical industry, 24
coal industry, 127, 151–2
coal-miners, 151
consumer expenditures, 133
copper production, 91–2, 101

double glazing, 166–70
dust in atmosphere, 81

efficiency, definition of, 60
  of fuel industries, 44, 89–91
  of heat engines, 63
  of cars, petrol and electric, 65
  of electricity generation, 43, 90
  historical trend fuel industries, 132
electricity,
  efficiency of generation, 44
  trend and projection of efficiency, 90
  use of primary fuels for, 130
  used for space heating, 131, 163–5
  investment for producing, 181
  usefulness of, 64–7
employment, in car trade, 206
  in U.K. industries, 118

energy, explanation of, 23
  conservation of, 59
  conversions, 59, 68–9
  and G.N.P., 115–19
  flows in atmosphere, 76–82
energy cost, 22
energy degradation, 60
energy ratio, food production, 33
Eskimo, 31

factor inputs, 26
food production, 33–7
  and fuel use, 49, 161
  and bread, 54
  projected fuel use, t-f scenario, 161
  projected fuel use, 1-g scenario, 199
fuel, definition of, 25
  energy content of, 42
fuel consumption, U.K. 1968, 42
  personal, 40, 44
  by sector, 45–54
  by final demand, 54
  by type of use, 67
  of world, 75
  direct by industries, 118
  U.K. (historical), 127–8
  for space heating, 129
  by households (historical), 129
  and household income, 134
fuel cost, 25–6
  units of, 43
  of commodities, 55–6
  of bread, 54
  of house, 51

fuel cost – continued
   of fuels, 55, 96
   of nuclear-power systems, 97–109
fuel industries, efficiencies of, 44
   projections of efficiencies of, 132
fuel use and productivity, 32, 36

gravitational energy, 24
gross national product (G.N.P.),
   and fuel use, 114–19
   growth in, 130
   and fuel crisis, 126

heat, 60–64
heat engine, 61
heat limit, 70–88
heat release (in U.K.), 86–7
house, fuel cost of, 51
   fuel used in heating, 53
   heating trends, 129
   insulation of, 166–70
   solar heating of, 194–6

ice cover, 79

Kelvin temperature scale, 63, 71
kilowatt, kilowatt-hour, kW, kWh, 37
kilowatt-hour-termal kWht, 43
kinetic energy, 24

manpower, in industry, 118
   in agriculture, 35

North Sea oil, 137–9
   and E.E.C., 148
   and b-a-u oil demand, 145
   and t-f oil demand, 175–178
   and l-g oil demand, 203
nuclear power, energy analysis of, 97–109
   rate of growth in U.S.A., 112
   maximum U.K. capacity, 143

oil, U.K. trade in, 140
ore grade, 91

power, 37
   rating of machines, 38
primary fuel, 41
   input to U.K., 42
productivity, and power, 37
   of coal-miners, 151–2, 178
   in food production, 34–5

recycling, 170

solar heating, 194–6
solar input, to U.K., 85–7
   to world, 72

thermodynamics, 58–64
   first law of, 58
   second law of, 58, 61, 63
transport and fuel use, 52
   projections of, 220–28
   electric and petrol cars, 65

unemployment, 205–8
uranium, use in S.G.H.W.R., 100
   fuel cost of, 102
   cut-off grade of, 102
   total resources of, 141

world fuel consumption, 75

# More about Penguins and Pelicans

*Penguinews*, which appears every month, contains details of all the new books issued by Penguins as they are published. From time to time it is supplemented by *Penguins in Print*, which is a complete list of all titles available. (There are some five thousand of these.)

A specimen copy of *Penguinews* will be sent to you free on request. For a year's issues (including the complete lists) please send 50p if you live in the British Isles, or 75p if you live elsewhere. Just write to Dept EP, Penguin Books Ltd, Harmondsworth, Middlesex, enclosing a cheque or postal order, and your name will be added to the mailing list.

*In the U.S.A.*: For a complete list of books available from Penguin in the United States write to Dept CS, Penguin Books Inc., 7110 Ambassador Road, Baltimore, Maryland 21207.

*In Canada*: For a complete list of books available from Penguin in Canada write to Penguin Books Canada Ltd, 41 Steelcase Road West, Markham, Ontario

# The Property Machine

*Peter Ambrose and Bob Colenutt*

A notice appears outside a row of terraced houses. *For development, 416,000 Square Feet of Prime Office Space.* Later twenty storeys of glass and concrete rear up into the atmosphere.

We have all seen it happen. Throughout Britain cities have been radically altered in outline and structure by a spate of office building. Peter Ambrose and Bob Colenutt spell out why this has happened, who allowed it to happen and what it actually means to the community.

Writing from a non-technical standpoint, they examine in detail the main components of the property system – the developers, the financial institutions, and the local and central planning machinery. They then look closely at what has happened in two specific areas, Brighton and Southwark, and conclude with a set of prescriptions for change. For, as they clearly reveal, office development often exacts a high price from the man in the street, as he faces soaring rents and rates, a scarcity of domestic housing and the enforced break-up of local communities.

# Thalidomide and the Power of the Drug Companies

*Henning Sjöström and Robert Nilsson*

Thalidomide... was this just an innocent case of a tranquillizer turning out to have monstrous side-effects on children to be born? Or was something uglier at work in its destructive career?

A Swedish lawyer and a research chemist have pieced together, for this Penguin Special, the story of the companies that made and sold the drug.

They argue that thalidomide was known to be dangerous (for the damage it could do to the nervous system) when it was put on the market; that the threat represented to children in the womb was recognized for some time before it was withdrawn.

Recalling the legal battles which were fought around the drug in Western Europe, the U.S.A., Japan and Australia, the authors (one of whom actively advised the prosecution in some countries) do not hesitate to name the scientists working for the companies and quote from their evidence and memoranda. Their story strongly suggests that the mysteries of science may place too much power in the hands of those who are out for profits, and in their final chapter they propose ways of preventing the tragic case of thalidomide being repeated.

*Not for sale in Scandinavia*